Civil Engineering: Construction Planning and Management

Civil Engineering: Construction Planning and Management

Edited by Jim Griffiths

CLANRYE INTERNATIONAL
www.clanryeinternational.com

Clanrye International,
750 Third Avenue, 9th Floor,
New York, NY 10017, USA

ISBN: 978-1-63240-712-2

Cataloging-in-Publication Data

Civil engineering : construction planning and management / edited by Jim Griffiths.
p. cm.
Includes bibliographical references and index.
ISBN 978-1-63240-712-2
1. Civil engineering. 2. Construction industry--Management. I. Griffiths, Jim.
TA145 .C58 2018
624--dc23

For information on all Clanrye International publications
visit our website at www.clanryeinternational.com

CLANRYE INTERNATIONAL

Printed in China.

Contents

Preface

Civil engineering is an interdisciplinary field concerned with the planning, construction and management of built environment. Construction planning and management refers to the process of designing and constructing any building, roads, bridges, etc. Its main purpose is to control and check the quality and cost of the project. The different types of construction that fall under this subject are institutional, agricultural, environmental, residential, heavy civil, industrial, etc. This text picks up individual branches and explains their need and contribution in the context of the growth of this field. The topics covered herein deal with the core aspects of the area. This textbook will serve as a reference to a broad spectrum of readers.

A foreword of all Chapters of the book is provided below:

Chapter 1 - Construction engineering is a sub-category of civil engineering. It focuses on construction, planning, and designing of roads, buildings and other related infrastructure. Civil engineering projects are divided into building projects, heavy engineering projects and industrial projects. The section on the fundamentals of construction engineering offers an insightful focus, keeping in mind the complex subject matter; **Chapter 2 -** A civil engineering project is shared among the owner, the designer and the contractor. Good communication among these members and various others involved is necessary for a successful project. Delivery methods or contractual relation and types of contract deal with the various forms of communication. The major categories of types of contracts and contract management are dealt with great details in the chapter; **Chapter 3 -** For any form of construction work, it is necessary to determine the cost to carry out and complete the project. The construction project manager estimates the quantity and cost. This is known as the estimate. Construction work estimate is classified into preliminary and detailed estimate. The topics discussed in the section are of great importance to broaden the existing knowledge on estimation concepts and rate analysis; **Chapter 4 -** Project management is essential to produce or develop a service or a product. Since a project is a temporary undertaking, normally time-bound, its management assumes necessity to achieve the desired goal. Project planning, a part of project management, uses schedules assess progress. This chapter has been carefully written to provide an easy understanding of the varied facets of project planning and management.

I would like to thank the entire editorial team who made sincere efforts for this book and my family who supported me in my efforts of working on this book. I take this opportunity to thank all those who have been a guiding force throughout my life.

Editor

1

Fundamentals of Construction Engineering

Construction engineering is a sub-category of civil engineering. It focuses on construction, planning, and designing of roads, buildings and other related infrastructure. Civil engineering projects are divided into building projects, heavy engineering projects and industrial projects. The section on the fundamentals of construction engineering offers an insightful focus, keeping in mind the complex subject matter.

Construction

In large construction projects, such as this skyscraper in Melbourne, Australia, cranes are essential.

Construction is the process of constructing a building or infrastructure. Construction differs from manufacturing in that manufacturing typically involves [[¥mass production]] of similar items without a designated purchaser, while construction typically takes place on location for a known client. Construction as an industry comprises six to nine percent of the gross domestic product of developed countries. Construction starts with planning, design, and financing; and continues until the project is built and ready for use.

Large-scale construction requires collaboration across multiple disciplines. An architect normally manages the job, and a construction manager, design engineer, construction engineer or project manager supervises it. For the successful execution of a project, effective planning is essential. Those involved with the design and execution of the infrastructure in question must consider zoning requirements, the environmental impact of the job, the successful scheduling, budgeting, construction-site safety, availability and transportation of building materials, logistics, inconvenience to the public caused by construction delays and bidding, etc. The largest construction projects are referred to as megaprojects.

Types of Construction

Industrial assemblage of a thermal oxidizer in the United States of America

Military residential unit construction by U.S. Navy personnel in Afghanistan

In general, there are three sectors of construction: buildings, infrastructure and industrial. Building construction is usually further divided into residential and non-residential (commercial/institutional). Infrastructure is often called heavy/highway, heavy civil or heavy engineering. It includes large public works, dams, bridges, highways, water/wastewater and utility distribution. Industrial includes refineries, process chemical, power generation, mills and manufacturing plants. There are other ways to break the industry into sectors or markets.

Engineering News-Record (ENR) is a trade magazine for the construction industry. Each year, ENR compiles and reports on data about the size of design and construction companies. They publish a list of the largest companies in the United States (Top-40) and also a list the largest global firms (Top-250, by amount of work they are doing outside their home country). In 2014, ENR compiled the data in nine market segments. It was divided as transportation, petroleum, buildings, power, industrial, water, manufacturing, sewer/waste, telecom, hazardous waste plus a tenth category for other projects. In their reporting on the Top 400, they used data on transportation, sewer, hazardous waste and water to rank firms as heavy contractors.

The Standard Industrial Classification and the newer North American Industry Classification System have a classification system for companies that perform or otherwise engage in construction. To recognize the differences of companies in this sector, it is divided into three subsectors: building construction, heavy and civil engineering construction, and specialty trade contractors. There are also categories for construction service firms (e.g., engineering, architecture) and construction managers (firms engaged in managing construction projects without assuming direct financial responsibility for completion of the construction project).

Building Construction

Building construction is the process of adding structure to real property or construction of buildings. The majority of building construction jobs are small renovations, such as addition of a room, or renovation of a bathroom. Often, the owner of the property acts as laborer, paymaster, and design team for the entire project. Although building construction projects typically include various common elements, such as design, financial, estimating and legal considerations, many projects of varying sizes reach undesirable end results, such as structural collapse, cost overruns, and/or litigation. For this reason, those with experience in the field make detailed plans and maintain careful oversight during the project to ensure a positive outcome.

The National Cement Share Company of Ethiopia's new plant in Dire Dawa.

Commercial building construction is procured privately or publicly utilizing various delivery methodologies, including cost estimating, hard bid, negotiated price, traditional, management contracting, construction management-at-risk, design & build and design-build bridging.

Residential construction practices, technologies, and resources must conform to local building authority regulations and codes of practice. Materials readily available in the area generally dictate the construction materials used (e.g. brick versus stone, versus timber). Cost of construction on a per square meter (or per square foot) basis for houses can vary dramatically based on site conditions, local regulations, economies of scale (custom designed homes are often more expensive to build) and the availability of skilled tradespeople. As residential construction (as well as all other types of construction) can generate a lot of waste, careful planning again is needed here.

Residential Construction

The most popular method of residential construction in North America is wood-framed construction. Typical construction steps for a single-family or small multi-family house are:

- Develop floor plans and obtain a materials list for estimations (more recently performed with estimating software)
- Obtain government building approval if necessary
- Clear the building site

- Survey to stake out for the foundation
- Excavate the foundation and dig footers.
- Pour a foundation and footers with concrete
- Build the main load-bearing structure out of thick pieces of wood and possibly metal I-beams for large spans with few supports.
- Add floor and ceiling joists and install subfloor panels
- Cover outer walls and roof in OSB or plywood and a water-resistive barrier.
- Install roof shingles or other covering for flat roof
- Cover the walls with siding, typically vinyl, wood, or brick veneer but possibly stone or other materials
- Install windows

Framing

- Frame interior walls with wooden 2x4s
- Add internal plumbing, HVAC, electrical, and natural gas utilities
- Building inspector visits if necessary to approve utilities and framing
- Install insulation and interior drywall panels (cementboard for wet areas) and to complete walls and ceilings
- Install bathroom fixtures
- Spackle, prime, and paint interior walls and ceilings

- Additional tiling on top of cementboard for wet areas, such as the bathroom and kitchen backsplash
- Install final floor covering, such as floor tile, carpet, or wood flooring
- Install major appliances
- Unless the original owners are building the house, at this point it is typically sold or rented.

New Construction Techniques and Sustainability

As efficiency codes have come into effect in recent years, new construction technologies and methods have emerged. University Construction Management departments are on the cutting edge of the newest methods of construction intended to improve efficiency, performance and reduce construction waste.

New techniques of building construction are being researched, made possible by advances in 3D printing technology. In a form of *additive building construction*, similar to the additive manufacturing techniques for manufactured parts, building printing is making it possible to flexibly construct small commercial buildings and private habitations in around 20 hours, with built-in plumbing and electrical facilities, in one continuous build, using large 3D printers. Working versions of 3D-printing building technology are already printing 2 metres (6 ft 7 in) of building material per hour as of January 2013, with the next-generation printers capable of 3.5 metres (11 ft) per hour, sufficient to complete a building in a week. Dutch architect Janjaap Ruijssenaars's performative architecture 3D-printed building was scheduled to be built in 2014.

In the current trend of sustainable construction, the recent movements of New Urbanism and New Classical Architecture promote a sustainable approach towards construction, that appreciates and develops smart growth, architectural tradition and classical design. This is in contrast to modernist and short-lived globally uniform architecture, as well as opposing solitary housing estates and suburban sprawl. Both trends started in the 1980s.

The construction site may be shut down due to bad weather. Erecting scaffolded tents over the site may reduce the number of lost work days, increasing productivity.

Construction Processes

Design Team

In the modern industrialized world, construction usually involves the translation of designs into reality. A formal design team may be assembled to plan the physical proceedings, and to integrate those proceedings with the other parts. The design usually consists of drawings and specifications, usually prepared by a design team including Architect, civil engineers, mechanical engineers, electrical engineers, structural engineers, fire protection engineers, planning consultants, architectural consultants, and archaeological consultants. The design team is most commonly employed by (i.e. in contract with) the property owner. Under this system, once the design is completed by the design team, a number of construction companies or construction management companies may then be asked to make a bid for the work, either based directly on the design, or on the basis of drawings and a bill of quantities provided by a quantity surveyor. Following evaluation of bids, the owner typically awards a contract to the most cost efficient bidder.

Shasta Dam under construction in June 1942

The best modern trend in design is toward integration of previously separated specialties, especially among large firms. In the past, architects, interior designers, engineers, developers, construction managers, and general contractors were more likely to be entirely separate companies, even in the larger firms. Presently, a firm that is nominally an "architecture" or "construction management" firm may have experts from all related fields as employees, or to have an associated company that provides each necessary skill. Thus, each such firm may offer itself as "one-stop shopping" for a construction project, from beginning to end. This is designated as a "design build" contract where the contractor is given a performance specification and must undertake the project from design to construction, while adhering to the performance specifications.

Several project structures can assist the owner in this integration, including design-build, partnering and construction management. In general, each of these project structures allows the owner to integrate the services of architects, interior designers, engineers and constructors throughout design and construction. In response, many companies are growing beyond traditional offerings of design or construction services alone and are placing more emphasis on establishing relationships with other necessary participants through the design-build process.

The increasing complexity of construction projects creates the need for design professionals trained in all phases of the project's life-cycle and develop an appreciation of the building as an advanced technological system requiring close integration of many sub-systems and their individual components, including sustainability. Building engineering is an emerging discipline that attempts to meet this new challenge.

Financial Advisors

Trump International Hotel and Tower (Chicago)

Construction projects can suffer from preventable financial problems. Underbids happen when builders ask for too little money to complete the project. Cash flow problems exist when the present amount of funding cannot cover the current costs for labour and materials, and because they are a matter of having sufficient funds at a specific time, can arise even when the overall total is enough. Fraud is a problem in many fields, but is notoriously prevalent in the construction field. Financial planning for the project is intended to ensure that a solid plan with adequate safeguards and contingency plans are in place before the project is started and is required to ensure that the plan is properly executed over the life of the project.

May 23, 2006

September 14, 2007 (3 months before completion)

Mortgage bankers, accountants, and cost engineers are likely participants in creating an overall plan for the financial management of the building construction project. The presence of the mortgage banker is highly likely, even in relatively small projects since the owner's equity in the property is the most obvious source of funding for a building project. Accountants act to study the expected monetary flow over the life of the project and to monitor the payouts throughout the process. Cost engineers and estimators apply expertise to relate the work and materials involved to a proper valuation. Cost overruns with government projects have occurred when the contractor identified change orders or project changes that increased costs, which are not subject to competition from other firms as they have already been eliminated from consideration after the initial bid.

Large projects can involve highly complex financial plans and often start with a conceptual estimate performed by a building estimator. As portions of a project are completed, they may be sold, supplanting one lender or owner for another, while the logistical requirements of having the right trades and materials available for each stage of the building construction project carries forward. In many English-speaking countries, but not the United States, projects typically use quantity surveyors.

Legal Aspects

Construction along Ontario Highway 401, widening the road from six to twelve travel lanes

Construction site with tubes

A construction project must fit into the legal framework governing the property. These include governmental regulations on the use of property, and obligations that are created in the process of construction.

The project must adhere to zoning and building code requirements. Constructing a project that fails to adhere to codes does not benefit the owner. Some legal requirements come from malum in se considerations, or the desire to prevent things that are indisputably bad – bridge collapses or explosions. Other legal requirements come from malum prohibitum considerations, or things that are a matter of custom or expectation, such as isolating businesses to a business district and residences to a residential district. An attorney may seek changes or exemptions in the law that governs the land where the building will be built, either by arguing that a rule is inapplicable (the bridge design will not cause a collapse), or that the custom is no longer needed (acceptance of live-work spaces has grown in the community).

A construction project is a complex net of contracts and other legal obligations, each of which all parties must carefully consider. A contract is the exchange of a set of obligations between two or more parties, but it is not so simple a matter as trying to get the other side to agree to as much as possible in exchange for as little as possible. The time element in construction means that a delay costs money, and in cases of bottlenecks, the delay can be extremely expensive. Thus, the contracts must be designed to ensure that each side is capable of performing the obligations set out. Contracts that set out clear expectations and clear paths to accomplishing those expectations are far more likely to result in the project flowing smoothly, whereas poorly drafted contracts lead to confusion and collapse.

Legal advisors in the beginning of a construction project seek to identify ambiguities and other potential sources of trouble in the contract structure, and to present options for preventing problems. Throughout the process of the project, they work to avoid and resolve conflicts that arise. In each case, the lawyer facilitates an exchange of obligations that matches the reality of the project.

Interaction of Expertise

Apartment complex under construction in Daegu, South Korea

Design, finance, and legal aspects overlap and interrelate. The design must be not only structurally sound and appropriate for the use and location, but must also be financially possible to build, and legal to use. The financial structure must accommodate the need for building the design provided, and must pay amounts that are legally owed. The legal structure must integrate the design into the surrounding legal framework, and enforce the financial consequences of the construction process.

Procurement

Procurement describes the merging of activities undertaken by the client to obtain a building. There are many different methods of construction procurement; however the three most common types of procurement are traditional (design-bid-build), design-build and management contracting.

There is also a growing number of new forms of procurement that involve relationship contracting where the emphasis is on a co-operative relationship between the principal and contractor and other stakeholders within a construction project. New forms include partnering such as Public-Private Partnering (PPPs) aka private finance initiatives (PFIs) and alliances such as "pure" or "project" alliances and "impure" or "strategic" alliances. The focus on co-operation is to ameliorate the many problems that arise from the often highly competitive and adversarial practices within the construction industry.

Traditional

This is the most common method of construction procurement and is well established and recognized. In this arrangement, the architect or engineer acts as the project coordinator. His or her role is to design the works, prepare the specifications and produce construction drawings, administer the contract, tender the works, and manage the works from inception to completion. There are direct contractual links between the architect's client and the main contractor. Any subcontractor has a direct contractual relationship with the main contractor. The procedure continues until the building is ready to occupy.

Design-build

Construction of the *Phase-1* (first two towers) of the Havelock City Project, Sri Lanka

This approach has become more common in recent years, and involves the client contracting a single entity to both provide a design and to build that design. In some cases, the design-build package can also include finding the site, arranging funding and applying for all necessary statutory consents.

The owner produces a list of requirements for a project, giving an overall view of the project's goals. Several D&B contractors present different ideas about how to accomplish these goals. The owner selects the ideas he or she likes best and hires the appropriate contractor. Often, it is not just one contractor, but a consortium of several contractors working together. Once these have been hired, they begin building the first phase of the project. As they build phase 1, they design phase 2. This is in contrast to a design-bid-build contract, where the project is completely designed by the owner, then bid on, then completed.

Kent Hansen pointed out that state departments of transportation usually use design build contracts as a way of progressing projects when states lack the skills-resources. In such departments, design build contracts are usually employed for very large projects.

Management Procurement Systems

In this arrangement the client plays an active role in the procurement system by entering into separate contracts with the designer (architect or engineer), the construction manager, and individual trade contractors. The client takes on the contractual role, while the construction or project manager provides the active role of managing the separate trade contracts, and ensuring that they complete all work smoothly and effectively together.

Management procurement systems are often used to speed up the procurement processes, allow the client greater flexibility in design variation throughout the contract, give the ability to appoint individual work contractors, separate contractual responsibility on each individual throughout the contract, and to provide greater client control.

In recent time, construction software starts to get traction - as it digitizes construction industry. Among solutions, there are for example: Procore, GenieBelt, PlanGrid, etc.

Authority Having Jurisdiction

In construction, the authority having jurisdiction (AHJ) is the governmental agency or sub-agency that regulates the construction process. In most cases, this is the municipality where the building is located. However, construction performed for supra-municipal authorities are usually regulated directly by the owning authority, which becomes the AHJ.

Construction on the Federal Reserve building in Kansas City, Missouri

Before the foundation can be dug, contractors are typically required to verify and have existing utility lines marked, either by the utilities themselves or through a company specializing in such services. This lessens the likelihood of damage to the existing electrical, water, sewage, phone, and cable facilities, which could cause outages and potentially hazardous situations. During the construction of a building, the municipal building inspector inspects the building periodically to ensure that the construction adheres to the approved plans and the local building code. Once construction is complete and a final inspection has been passed, an occupancy permit may be issued.

An operating building must remain in compliance with the fire code. The fire code is enforced by the local fire department or a municipal code enforcement office.

Changes made to a building that affect safety, including its use, expansion, structural integrity, and fire protection items, usually require approval of the AHJ for review concerning the building code.

Industry Characteristics

In the United States, the industry in 2014 has around $960 billion in annual revenue according to statistics tracked by the Census Bureau, of which $680 billion is private (split evenly between residential and nonresidential) and the remainder is government. As of 2005, there were about 667,000 firms employing 1 million contractors (200,000 general contractors, 38,000 heavy, and 432,000 specialty); the average contractor employed fewer than 10 employees. As a whole, the industry employed an estimated 5.8 million as of April 2013, with a 13.2% unemployment rate. In the United States, approximately 828,000 women were employed in the construction industry as of 2011.

Careers

Helicopter view of the Atacama Large Millimeter/submillimeter Array (ALMA) Operations Support Facility (OSF) construction site

Ironworkers erecting the steel frame of a new building at Massachusetts General Hospital in Boston

Skilled occupations include carpenters, electricians, plumbers, ironworkers, masons, and many other manual crafts, as well as those involved in project management. In the UK these require further education qualifications, often in vocational subject areas. These qualifications are either obtained directly after the completion of compulsory education or through "on the job" apprenticeship training. In the UK, 8500 construction-related apprenticeships were commenced in 2007.

A truck operator at Al Gamil, the largest construction company in Djibouti.

There are many routes to the different careers within the construction industry. These three main tiers are based on educational background and training, which vary by country:

- Unskilled and semi-skilled – General site labor with little or no construction qualifications.
- Skilled – Tradesmen who've served apprenticeships, typically in labor unions, and on-site managers who possess extensive knowledge and experience in their craft or profession.
- Technical and management – Personnel with the greatest educational qualifications, usually graduate degrees, trained to design, manage and instruct the construction process.

Technical and specialized occupations require more training as a greater technical knowledge is required. These professions also hold more legal responsibility. A short list of the main careers with an outline of the educational requirements are given below:

- Architect – Typically holds 1, undergraduate 3 year degree in architecture + 1, post-graduate 2 year degree (DipArch or BArch) in architecture plus 24 months experience within the industry. To use the title «architect» the individual must be registered on the Architects Registration Board register of Architects.
- Civil engineer – Typically holds a degree in a related subject. The Chartered Engineer qualification is controlled by the Engineering Council, and is often achieved through membership of the Institution of Civil Engineers. A new university graduate must hold a master's degree to become chartered; persons with bachelor's degrees may become an Incorporated Engineer.
- Building services engineer – Often referred to as an «M&E Engineer» typically holds a degree in mechanical or electrical engineering. Chartered Engineer status is governed by the Engineering Council, mainly through the Chartered Institution of Building Services Engineers.
- Project manager – Typically holds a 4-year or greater higher education qualification, but are often also qualified in another field such as architecture, civil engineering or quantity surveying.

- Structural engineer – Typically holds a bachelor›s or master›s degree in structural engineering. A P.ENG is required from the Professional Engineers Ontario (Canada). New university graduates must hold a master›s degree to gain chartered status from the Engineering Council, mainly through the Institution of Structural Engineers (UK).
- Quantity surveyor – Typically holds a bachelor›s degree in quantity surveying. Chartered status is gained from the Royal Institution of Chartered Surveyors.
- Civil estimators are professionals who typically have a background in civil engineering, construction project management, or construction supervision.

In 2010 a salary survey revealed the differences in remuneration between different roles, sectors and locations in the construction and built environment industry. The results showed that areas of particularly strong growth in the construction industry, such as the Middle East, yield higher average salaries than in the UK for example. The average earning for a professional in the construction industry in the Middle East, across all sectors, job types and levels of experience, is £42,090, compared to £26,719 in the UK. This trend is not necessarily due to the fact that more affluent roles are available, however, as architects with 14 or more years experience working in the Middle East earn on average £43,389 per annum, compared to £40,000 in the UK. Some construction workers in the US/Canada have made more than $100,000 annually, depending on their trade.

Safety

At-risk workers without appropriate safety equipment

Construction is one of the most dangerous occupations in the world, incurring more occupational fatalities than any other sector in both the United States and in the European Union. In 2009, the fatal occupational injury rate among construction workers in the United States was nearly three times that for all workers. Falls are one of the most common causes of fatal and non-fatal injuries among construction workers. Proper safety equipment such as harnesses and guardrails and procedures such as securing ladders and inspecting scaffolding can curtail the risk of occupational injuries in the construction industry. Other major causes of fatalities in the construction industry include electrocution, transportation accidents, and trench cave-ins.

Roofing requires a very high level of safety

Other safety risks for workers in construction include hearing loss due to high noise exposure, musculoskeletal injury, chemical exposure, and high levels of stress.

History

A picture of a building under construction in India.

The first huts and shelters were constructed by hand or with simple tools. As cities grew during the Bronze Age, a class of professional craftsmen, like bricklayers and carpenters, appeared. Occasionally, slaves were used for construction work. In the Middle Ages, these were organized into guilds. In the 19th century, steam-powered machinery appeared, and later diesel- and electric powered vehicles such as cranes, excavators and bulldozers.

Fast-track construction has been increasingly popular in the 21st century. Some estimates suggest that 40% of construction projects are now fast-track construction.

Construction Phases

- Vision/fantasy/idea - a concept never intended to be built, may be an aesthetic or structural design exercise
- Proposed - a building concept that is under review by the building owner and by government
- Approved - a building concept that will be constructed in the near future. If the proposed building is not approved then the proposal may be amended and resubmitted, or it may be deferred or cancelled.
- Design - the specification of what is to be built in sufficient detail to be used as the basis as a contract between the owner and a contractor
- Procurement - the selection of the contractor or contractors to carry out the construction. This may be by competitive tendering.
- Diversions - before construction can start any services on the site which must be kept operational to serve other adjacent sites must be diverted so they run outside the footprint of the new building. This can include drainage, water and gas piped services and power and communication cables.
- Under-construction - a fully designed building currently being built
- Ground works - construction work below ground level including the construction of basements and foundations
- Topped-out - a fully designed building where construction has reached the highest point of the building
- Fitting out - installation of the decorative, non-structural elements once the building main structure is complete. This includes painting, ceilings, light fittings etc.
- Commissioning or setting to work - Once the building Mechanical, electrical, plumbing, communications, and building control systems are installed they then need to be tested and adjusted so they deliver the required performance. In modern buildings this can take some time during which little seems to be going on but if this is not done properly then these systems will not deliver their design performance leading to hot and cold spots, spurious alarms, higher energy bills, and systems failing during emergencies
- Substantial Completion / Beneficial Occupancy - a point when the work is sufficiently complete so that the Owner can occupy (Items noted during inspection 'punch list' or 'snag list' may still be corrected)
- Complete/built - a fully designed building that has been fully built, excluding future expansions (punch list items all completed)
- Building Operation - All those day-to-day activities need to ensure the building can be used. In simple buildings this means little more than cleaning but in more complicated buildings this is a large scale operation employing a large team of staff. If they do their job right then you hardly notice them.

- Maintenance - works to ensure the building continues to operate in accordance with its design, including replacing elements which are approaching the end of their useful life
- Repair - replacing building elements which have been damaged or which have failed to restore the building to its as-built state
- Renovation - modification to the building. This can be minor modifications that are carried out while the building is occupied or major works where only the structural elements are kept and the building is out of use for years
- Demolition - destruction of the building which may include the salvage of some elements for reuse elsewhere.

Construction Engineering

Construction engineering is a professional discipline that deals with the designing, planning, construction, and management of infrastructures such as roads, tunnels, bridges, airports, railroads, facilities, buildings, dams, utilities and other projects. Construction engineering is considered a professional sub-practice area of Civil engineering.

Construction technology is a related field that deals more with the practical aspects of projects. Construction technologists or construction technicians learn some of the design aspects similar to civil engineers and some of the project site management aspects similar to construction managers. These technicians are unique such that they are a cross between civil engineers and construction managers.

At the educational level, civil engineering students concentrate primarily on the design work, which is more analytical, gearing them toward a career as a design professional. This essentially requires them to take a multitude of challenging engineering science and design courses as part of obtaining a 4 year accredited degree. Also, education for construction managers is primarily focused on construction procedures, methods, costs, schedules, and personnel management. Their primary concern is to deliver a project on time, within budget, and of the desired quality.

The difference between a construction technologist and a civil engineer is that civil engineering is an engineering discipline. Construction technology students take basic design courses as well as construction management courses. This allows them to understand both basic design functions as well as the building requirements needed to design and build today's infrastructures.

Work Activities

Depending on which career the construction engineer has chosen to follow, an entry-level design engineer normally provides support to project managers and assist with creating conceptual designs, scopes, and cost estimates for the planning and construction of approved projects. It should be noted that a career in design work does require a professional engineer license (PE). Individuals who pursue this career path are strongly advised to sit for the Engineer In Training exam (EIT) while in college as it takes five years (4 years in USA) post graduate to obtain the PE license.

Entry-level construction manager positions are typically called project engineers or assistant project engineers. They are responsible for preparing purchasing requisitions, processing change orders, preparing monthly budgeting reports, and handling meeting minutes. The construction management position does not necessarily require a PE license; however possessing one does make the individual more marketable, as the PE license allows the individual to sign off on temporary structure designs.

Abilities

Construction engineers are problem solvers. They contribute to the creation of infrastructure that best meets the unique demands of its environment. They must be able to understand infrastructure life cycles. When compared and contrasted to design engineers, construction engineers bring to the table their own unique perspectives for solving technical challenges with clarity and imagination. While individuals considering this career path should certainly have a strong understanding of mathematics and science, many other skills are also highly desirable, including critical and analytical thinking, time management, people management, and good communication skills.

Educational Requirements

Individuals looking to obtain a construction engineering degree must first ensure that the program is accredited by EAC or Technology Accreditation Commission (TAC) of the Accreditation Board for Engineering and Technology (ABET). ABET accreditation is assurance that a college or university program meets the quality standards established by the profession for which it prepares its students. In the US there are currently twenty-five programs that exist in the entire country so careful college consideration is advised.

A typical construction engineering curriculum is a mixture of engineering mechanics, engineering design, construction management and general science and mathematics. This usually leads to a Bachelor of Science degree. The B.S. degree along with some design or construction experience is sufficient for most entry level positions. Graduate schools may be an option for those who want to go further in depth of the construction and engineering subjects taught at the undergraduate level. In most cases construction engineering graduates look to either civil engineering, engineering management, or business administration as a possible graduate degree.

Job Prospects

Job prospects for construction engineers generally have a strong cyclical variation. For example, starting in 2008 - continuing until at least 2011 - job prospects have been poor due to the collapse of housing bubbles in many parts of the world. This sharply reduced demand for construction, forced construction professionals towards infrastructure construction and therefore increased the competition faced by established and new construction engineers. This increased competition, and a core reduction in quantity demand is in parallel with a possible shift in the demand for construction engineers due to the automation of many engineering tasks, overall resulting in reduced prospects for construction engineers. In early 2010 the United States construction industry had a 27% unemployment rate, this is nearly three times higher than the 9.7% national average unemployment rate. The construction unemployment rate (including tradesmen) is comparable to the United States 1933 unemployment rate - the lowest point of the Great Depression - of 25%.

Remuneration

The average salary for a civil engineer in the UK depends on the sector, and more specifically the level of experience of the individual. A 2010 survey of the remuneration and benefits of those occupying jobs in construction and the built environment industry showed that the average salary of a civil engineer in the UK is £29,582. In the United States, as of May 2013, the average was $85,640. The average salary varies depending on experience, for example the average annual salary for a civil engineer with between 3 and 6 years experience is £23,813. For those with between 14 and 20 years experience the average is £38,214.

The art of construction has slowly developed into a science over the centuries. Civil engineers are involved from conception to physical realization of the project. We shall first enumerate the different civil engineering projects.

Building Project

1. Residential Building, Town Planning

Govt Staff Quaters	Residential Buildings		

2. Other Buildings

Academic building such as institutes, university and schools, hospitals, temples, government and private offices, commercial complexes, recreational building such as auditorium, cinema hall, theaters, stores, warehouse, light manufacturing plants.

Guest House	School	Hospital	Temple

Depending upon the level of development of a country, these construction projects constitute large share. Developed countries have more facilities in this area than porly developed countries.

Transportation Systems

Bridges and flyover, roads, railways, airport, urban mass transit system such as mono railway and ropeways , town planning, port and harbour, tunnel.

Bridges & Flyover	Airport	Road

These are basic infrastructural projects. Developing countries are spending major of its budget on these projects.

Hydraulic Structures

Dam, barrage, canal, hydroelectric power plant, tunnel, embankment, flood control system.

Dam	Canal	Barrage

For a country whose economy is dependent on agriculture, these are major infrastructural projects. India spent almost 1/5 th of its planning budget on these projects in first and second five year plan.

Fossil Fuel and Nuclear Power Plants

Nuclear Plant

Thermal Power Plants

Energy is vital for survival. The generation of electricity is carried out from coal, petroleum and nuclear fuel.

Offshore Stuctures

Tension Legged Platform

These are huge steel and concrete structure in sea, used usually for petroleum extraction.

Space Based and Defense Structure

Army Cantonment, missile launch pad, bunkers

Army Cantonment

Missle Launch Pad

Defence is an important part of any country. Thses are huge civil engineering construction projects related to defence. Technology involved is very sophisticated.

Industrial Structures

Petrochemical refinaries, small and large scale manufacturing plants, silos, bunker, mine developments.

Refinary

These are the backbone of modern growth of a country.

Communication System

Transmission line tower, communication network, TV Tower.

Transmission Line

TV Tower

These structures are lifeline for present activities.

Water and Waste Project

- Water treatment plant
- Water distribution system - pipe, water tanks
- Sewage and storm water collection system

- Sewage treatment plants
- Sewage disposal system

Water Treatment	Waste Water Treatment	Effulent Treatment

These structures are important for environment friendly living and healthy life.

Special Projects

High altitude construction project, marine exploration project.

High altitude construction

These are used for scientific exploration, strategic importance.

Classification of Civil Engineering Projects

Civil Engineer Project manager or construction manager has to interact with various field of specializations depending upon the type of the project.

A project manager involved in the construction of hospital has to interact with medical professional.

Similarly, the construction of petroleum, refinery involves the complex interaction with chemical engineer, mechanical engineer, electrical engineers.

Based on the intersection of knowledge and interaction with other disciplines, civil engineering projects can be grouped as follows:

- Building Projects
- Heavy engineering Project
- Industrial Projects

Building construction projects can be further divided into Residential Buildings and commercial buildings. We shall discuss the nature of the project next :

Building Construction

During design phase, building design is coordinated by architect who works with other engineering disciplines such as structure engineer, electrical engineer whereas construction phase is coordinated by the civil engineers. But group leader in the building projects is generally architect. Building project can further be classified into

- Residential Building
- Commercial and other buildings.

Residential buildings in India are primarily single or double storey building. Multi-storied residential buildings are also getting constructed. Other building project produces variety of building ranging from hospital, temples, and small stores to commercial office tower, theatres and cinema hall, government buildings, recreational facilities, small schools to new institute like IIT or university.

Economically building project accounts for around 60 to 70 percent of construction project in a country. But most of the constructions are carried out in an unorganized way. In India, building construction projects are very fragmented and majority of construction in rural and semi urban area is carried out by non-engineering ways. Building construction in India is highly labour –intensive. Most of the labourers employed for building projects are semi-skilled and untrained. Uses of heavy equipments are very rare. Building projects are governed by many local as well as general laws, such as municipal laws, building laws etc.

Building

A building or edifice is a structure with a roof and walls standing more or less permanently in one place, such as a house or factory. Buildings come in a variety of sizes, shapes and functions, and have been adapted throughout history for a wide number of factors, from building materials available, to weather conditions, to land prices, ground conditions, specific uses and aesthetic reasons. To better understand the term *building* compare the list of nonbuilding structures.

Buildings serve several needs of society – primarily as shelter from weather, security, living space, privacy, to store belongings, and to comfortably live and work. A building as a shelter represents a physical division of the human habitat (a place of comfort and safety) and the *outside* (a place that at times may be harsh and harmful).

The Hôtel de Ville, building in Paris, France

A Shaolin Monastery building in China

Ever since the first cave paintings, buildings have also become objects or canvasses of much artistic expression. In recent years, interest in sustainable planning and building practices has also become an intentional part of the design process of many new buildings.

Definitions

A building and skybridge in Munich, Germany

The Rotunda in Hellerup, Denmark, a cylindric building made in steel frame and aluminum

Example of a religious building: the Great Mosque of Kairouan (also called the Mosque of Uqba), founded in 670, dates in its present state principally from the 9th century. The Great Mosque of Kairouan is located in the city of Kairouan, Tunisia.

Da Vinci building (at the left), in Mendoza, Argentina. Considered one of the 1000 architectures of the Americas.

The word *building* is both a noun and a verb also an adverb: the structure itself and the act of making it. As a noun, a building is 'a structure that has a roof and walls and stands more or less permanently in one place'; "there was a three-storey building on the corner"; "it was an imposing edifice". In the broadest interpretation a fence or wall is a building. However, the word *structure* is used more broadly than *building* including natural and man-made formations and does not necessarily have walls. Structure is more likely to be used for a fence. Sturgis' Dictionary included that "[building] differs from architecture in excluding all idea of artistic treatment; and it differs from construction in the idea of excluding scientific or highly skilful treatment." As a verb, building is the act of construction.

Structural height in technical usage is the height to the highest architectural detail on building from street-level. Depending on how they are classified, spires and masts may or may not be included in this height. Spires and masts used as antennas are not generally included. The definition of a *low-rise vs. a high-rise* building is a matter of debate, but generally three storeys or less is considered low-rise.

History

A report by Shinichi Fujimura of a shelter built 500 000 years ago is doubtful since Fujimura was later found to have faked many of his findings. Supposed remains of huts found at the Terra Amata site in Nice purportedly dating from 200 000 to 400 000 years ago have also been called into question. There is clear evidence of homebuilding from around 18 000 BC. Buildings became common during the Neolithic.

Types

A timber-framed house in Marburg, Germany

A block of tenements (apartments) in Bruntsfield, Edinburgh, Scotland

Residential

Single-family residential buildings are most often called houses or homes. Residential buildings containing more than one dwelling unit are called a duplex, apartment building to differentiate them from 'individual' houses. A condominium is an apartment that the occupant owns rather than rents. Houses may also be built in pairs (semi-detached), in terraces where all but two of the houses have others either side; apartments may be built round courtyards or as rectangular blocks surrounded by a piece of ground of varying sizes. Houses which were built as a single dwelling may later be divided into apartments or bedsitters; they may also be converted to another use e.g. an office or a shop.

Building types may range from huts to multimillion-dollar high-rise apartment blocks able to house thousands of people. Increasing settlement density in buildings (and smaller distances between buildings) is usually a response to high ground prices resulting from many people wanting to live close to work or similar attractors. Other common building materials are brick, concrete or combinations of either of these with stone.

Residential buildings have different names for their use depending if they are seasonal include holiday cottage (vacation home) or timeshare; size such as a cottage or great house; value such as a shack or mansion; manner of construction such as a log home or mobile home; proximity to the ground such as earth sheltered house, stilt house, or tree house. Also if the residents are in need of special care such as a nursing home, orphanage or prison; or in group housing like barracks or dormitories.

Historically many people lived in communal buildings called longhouses, smaller dwellings called pit-houses and houses combined with barns sometimes called housebarns.

Buildings are defined to be substantial, permanent structures so other dwelling forms such as houseboats, yurts, and motorhomes are dwellings but not buildings.

Multi-storey

A multi-storey is a building that has multiple floors.

Complex

Sometimes a group of inter-related (and possibly inter-connected) builds are referred to as a complex - for example a *housing complex, educational complex, hospital complex*, etc.

Creation

The practice of designing, constructing, and operating buildings is most usually a collective effort of different groups of professionals and trades. Depending on the size, complexity, and purpose of a particular building project, the project team may include:

- A real estate developer who secures funding for the project;
- One or more financial institutions or other investors that provide the funding
- Local planning and code authorities
- A Surveyor who performs an ALTA/ACSM and construction surveys throughout the project;

- Construction managers who coordinate the effort of different groups of project participants;
- Licensed architects and engineers who provide building design and prepare construction documents;
- The principal design Engineering disciplines which would normally include the following professionals:- Civil, Structural, Mechanical building services or HVAC (heating Ventilation and Air Conditioning) Electrical Building Services, Plumbing and drainage. Also other possible design Engineer specialists may be involved such as Fire (prevention), Acoustic, facade engineers,building physics,Telecomms, AV (Audio Visual), BMS (Building Management Systems)Automatic controls etc. These design Engineers also prepare construction documents which are issued to specialist contractors to obtain a price for the works and to follow for the installations.
- Landscape architects;
- Interior designers;
- Other consultants;
- Contractors who provide construction services and install building systems such as climate control, electrical, plumbing, Decoration, fire protection, security and telecommunications;
- Marketing or leasing agents;
- Facility managers who are responsible for operating the building.

Regardless of their size or intended use, all buildings in the US must comply with zoning ordinances, building codes and other regulations such as fire codes, life safety codes and related standards.

Vehicles—such as trailers, caravans, ships and passenger aircraft—are treated as “buildings” for life safety purposes.

Ownership and Funding

- Mortgage loan
- Real estate developer

Building Services

Physical Plant

Any building requires a certain amount of internal infrastructure to function, which includes such elements like heating / cooling, power and telecommunications, water and wastewater etc. Especially in commercial buildings (such as offices or factories), these can be extremely intricate systems taking up large amounts of space (sometimes located in separate areas or double floors / false ceilings) and constitute a big part of the regular maintenance required.

The BB&T Building in Macon, Georgia is constructed of aluminum.

Conveying Systems

Systems for transport of people within buildings:

- Elevator
- Escalator
- Moving sidewalk (horizontal and inclined)

Systems for transport of people between interconnected buildings:

- Skyway
- Underground city

Building Damage

A building in Massueville, Quebec, Canada engulfed by fire

Buildings may be damaged during the construction of the building or during maintenance. There are several other reasons behind building damage like accidents such as storms, explosions and subsidence caused by mining or poor foundations. Buildings also may suffer from fire damage and flooding in special circumstances. They may also become dilapidated through lack of proper maintenance or alteration work improperly carried out.

Tower Block

A high-rise residential apartment building in Tseung Kwan O, Hong Kong

A tower block, high-rise, apartment tower, residential tower, apartment block, block of flats, or office tower is a tall building or structure used as a residential and/or office building. In some areas it may be referred to as an "MDU", standing for "Multi Dwelling Unit". In the United States, such a structure is referred to as an apartment building or office building, while a group of such buildings is called an apartment complex or office complex.

One of a pair of 47-story high-rise residential complexes, a ubiquitous feature of the borough of Fort Lee, Bergen County, New Jersey, United States, in the New York City Metropolitan Area, with several over 90 metres (300 ft) tall.

There are numerous high-rise apartments in Navi Mumbai, India

High-rise buildings became possible with the invention of the elevator (lift) and cheaper, more abundant building materials. The materials used for the structural system of high-rise buildings are reinforced concrete and steel. Most North American style skyscrapers have a steel frame, while residential blocks are usually constructed of concrete. There is no clear difference between a tower block and a skyscraper, although a building with fifty or more stories is generally considered a skyscraper.

10 story high-rise in Manchester, New Hampshire, U.S.

High-rise structures pose particular design challenges for structural and geotechnical engineers, particularly if situated in a seismically active region or if the underlying soils have geotechnical risk factors such as high compressibility or bay mud. They also pose serious challenges to firefighters during emergencies in high-rise structures. New and old building design, building systems like the building standpipe system, HVAC systems (heating, ventilation and air conditioning), fire sprinkler system and other things like stairwell and elevator evacuations pose significant problems. Studies are often required to ensure that pedestrian wind comfort and wind danger concerns are addressed. In order to allow less wind exposure, to transmit more daylight to the ground and to appear more slender, many high-rises have a design with setbacks.

Apartment buildings have technical and economic advantages in areas of high population density, and have become a distinctive feature of housing accommodation in virtually all densely populated urban areas around the world. In contrast with low-rise and single-family houses, apartment blocks accommodate more inhabitants per unit of area of land and decrease the cost of municipal infrastructure.

16-floor Rusland Court tower block in Charlestown, Greater Manchester, United Kingdom.

Definition

Various bodies have defined “high-rise”:

- Emporis Standards defines a high-rise as “A multi-story structure between 35–100 meters tall, or a building of unknown height from 12–39 floors.”
- According to the building code of Hyderabad, India, a high-rise building is one with four floors or more, or 15 to 18 meters or more in height.
- The *New Shorter Oxford English Dictionary* defines a high-rise as “a building having many storeys”.
- The *International Conference on Fire Safety in High-Rise Buildings* defined a high-rise as “any structure where the height can have a serious impact on evacuation”
- In the U.S., the National Fire Protection Association defines a high-rise as being higher than 75 feet (23 meters), or about 7 stories.
- Most building engineers, inspectors, architects and similar professionals define a high-rise as a building that is at least 75 feet (23 m) tall.

History

High-rise apartment buildings had already appeared in antiquity: the *insulae* in ancient Rome and several other cities in the Roman Empire, some of which might have reached up to ten or more stories, one reportedly having 200 stairs. Because of the destruction caused by poorly built high-rise *insulae* collapsing, several Roman emperors, beginning with Augustus (r. 30 BC – 14 AD), set limits of 20–25 meters for multi-story buildings, but met with limited success, as these limits were often ignored despite the likelihood of taller *insulae* collapsing. The lower floors were typically occupied by either shops or wealthy families, while the upper stories were rented out to the lower classes. Surviving Oxyrhynchus Papyri indicate that seven-story buildings even existed in provincial towns, such as in third century AD Hermopolis in Roman Egypt.

These tower blocks were built in Shibam, Yemen in the 16th century, and are the tallest mudbrick buildings in the world

In Arab Egypt, the initial capital city of Fustat housed many high-rise residential buildings, some seven stories tall that could reportedly accommodate hundreds of people. Al-Muqaddasi, in the 10th century, described them as resembling minarets, while Nasir Khusraw, in the early 11th century, described some of them rising up to 14 stories, with roof gardens on the top story complete with ox-drawn water wheels for irrigating them. By the 16th century, Cairo also had high-rise apartment buildings where the two lower floors were for commercial and storage purposes and the multiple stories above them were rented out to tenants.

The skyline of many important medieval cities was dominated by large numbers of high-rising urban towers, which fulfilled defensive but also representative purposes. The residential Towers of Bologna numbered between 80 and 100 at a time, the largest of which still rise to 97.2 m. In Florence, a law of 1251 decreed that all urban buildings should be reduced to a height of less than 26 m, the regulation immediately put into effect. Even medium-sized towns such as San Gimignano are known to have featured 72 towers up to 51 m in height.

The Hakka people in southern China have adopted communal living structures designed to be easily defensible in the forms of Weilongwu and Tulou, the latter are large, enclosed and fortified earth building, between three and five stories high and housing up to 80 families. The oldest still standing tulou dates back from the 14th century.

Tower blocks were built in the Yemeni city of Shibam in the 16th century. The houses of Shibam are all made out of mud bricks, but about five hundred of them are tower houses, which rise five to sixteen stories high, with each floor having one or two apartments. This technique of building was implemented to protect residents from Bedouin attacks. While Shibam has existed for around two thousand years, most of the city's houses date from the 16th century. The city has the tallest mud buildings in the world, some more than 30 meters (100 feet) high. Shibam has been called "one of the oldest and best examples of urban planning based on the principle of vertical construction" or "Manhattan of the desert".

Currently, the tallest high-rise apartment building in the world is Chicago's John Hancock Center, constructed under the supervision of Skidmore, Owings & Merrill and completed in 1969. The building has 100 stories and stands 344 meters tall.

Modern Development

A house in Steinfurt, Westphalia, Germany, forming a "Y"

The residential tower block with its typical concrete construction are a familiar feature of Modernist architecture. Influential examples include Le Corbusier's "housing unit" his *Unité d'Habitation*, repeated in various European cities starting with his *Cité radieuse* in Marseille (1947–52), constructed of *béton brut*, rough-cast concrete, as steel for framework was unavailable in post-war France. Residential tower blocks became standard in housing urban populations displaced by slum clearances and "urban renewal".

Highrise projects after World War II typically rejected the classical designs of the early skyscrapers, instead embracing the uniform international style; many older skyscrapers were redesigned to suit contemporary tastes or even got demolished - such as New York's Singer Building, once the world's tallest skyscraper. However, with the movements of Postmodernism, New Urbanism and New Classical Architecture, that established since the 1980s, a more classical approach came back to global skyscraper design, that is popular today.

Other contemporary styles and movements in highrise design include organic, sustainable, neo-futurist, structuralist, high-tech, deconstructivist, blob, digital, streamline, novelty, critical regionalist, vernacular, Neo Art Deco and neo-historist, also known as revivalist.

Europe

Eastern Europe

Winter in Nowa Huta (Kraków), Poland

Typical 11-storey large-panel system building in Budapest-Kispest (built by the BHK III.)

Although some Eastern European countries during the interwar period, such as the Second Polish Republic, already started building housing estates that were considered to be of a very high standard for their time, many of these structures perished during the Second World War. In the Eastern Bloc, tower blocks were constructed in great numbers to produce plenty of cheap accommodation for the growing postwar populations of the USSR and its satellite states. This took place mostly in the 50s, 60s and 70s of the 20th century, though in the People's Republic of Poland this process started even earlier due to the severe damages that Polish cities sustained during World War II. Throughout the former Eastern Bloc countries, tower blocks built during the Soviet years make up much of the current housing estates and most of them were built in the specific socialist realist style of architecture that was dominant in the territories east of the Iron Curtain. In Romania, mass construction of standardized apartment blocks occurred under the systematization programme of Nicolae Ceauşescu, who, after a 1971 visit in North Korea, was impressed by the Juche ideology of that country. Systematization consisted largely of the demolition and reconstruction of existing villages, towns, and cities, in whole or in part, in order to build blocks of flats (*blocuri*). In the Czech Republic and Slovakia (then Czechoslovakia), panelák building under communism resulted from two main factors: the postwar housing shortage and the ideology of communist Czechoslovak leaders.

Painted paneláks in Prague, Czech Republic. In recent years, several Eastern European countries have started projects of colouring and beautifying existing communist blocks, so that they are no longer gray.

Nowadays, in Eastern European countries, opinions about these buildings vary greatly, with some deeming them as eyesores on their city's landscape while others glorify them as relics of a bygone age. Since the dissolution of the Soviet Union, and especially in the late 1990s and early 2000s, many of the former Eastern Bloc countries have begun construction of new, more expensive and modern housing.

The Śródmieście borough of Warsaw, the capital of Poland, has seen the development of an array of skyscrapers. Russia is also currently undergoing a dramatic buildout, growing a commercially shaped skyline. Moreover, the ongoing changes made to postwar housing estates in Eastern Europe since the 2000s in countries like Russia, Poland, parts of Germany, Czech Republic, and Estonia vary - ranging from simply applying a new coat of paint to the previously grey exterior to thorough modernization of entire buildings. In the European Union, among former communist countries, a majority of the population lives in flats in Latvia (65.1%), Estonia (63.8%), Lithuania (58.4%), Czech Republic (52.8%) and Slovakia (50.3%) (as of 2014, data from Eurostat). However, not all flat dwellers in Eastern Europe live in communist era blocks of flats; many live in buildings constructed after the fall of communism, and some in buildings surviving from the era before communism.

Great Britain

The three tower blocks of the Crossways Estate in Bow, London, United Kingdom, before their refurbishment

Tower blocks were built in the UK after the Second World War. The first residential tower block, “The Lawn” was constructed in Harlow, Essex in 1951; it is now a Grade II listed building. In many cases tower blocks were seen as a “quick-fix” to cure problems caused by crumbling and unsanitary 19th-century dwellings or to replace buildings destroyed by German aerial bombing. It was argued that towers surrounded by public open space could provide the same population density as the terraced housing and small private gardens they replaced, offering larger rooms and improved views whilst being cheaper to build. Initially, they were welcomed, and their excellent views made them popular living places. Later, as the buildings themselves deteriorated, they grew a reputation for being undesirable low cost housing, and many tower blocks saw rising crime levels, increasing their unpopularity. One response to this was the great increase in the number of housing estates built, which in turn brings its own problems. In the UK, tower blocks particularly lost popularity after the partial collapse of Ronan Point in east London in 1968.

Glasgow, the largest city in Scotland, is believed to contain the highest concentration of tower blocks in the UK: examples include the Hutchesontown C blocks in the Gorbals and the 20-storey blocks in Sighthill. However, on the whole, London has the largest number of high-rise residential buildings in the UK.

The Post-war British Tower Block Vision

Post-war Britain was the stage for a tower block "building boom"; from the 1950s to the late 1970s there was a dramatic increase in tower block construction. During this time, local authorities desired to impress their voters by building futuristic and imposing tower blocks, which would signify post-war progress. Both Patrick Dunleavy and Lynsey Hanley agree that architects and planners were influenced by Le Corbusier's promotion of high-rise architecture. The modern tower blocks were to include features that would foster desired forms of resident interaction, an example being the inclusion of Le Corbusier's streets in the sky in some estates.

As well as inspiring residents, local authority planners believed that the way tower blocks were constructed would save money. Generally, the tower blocks were built on cheap greenfield land skirting established cities. Although the property prices for these periphery sites were markedly cheaper than their inner city counterparts, they often had little access to public amenities, such as public transport. It was thought that the implementation of industrialised building techniques would lower costs too, as similar tower blocks would be replicated over many sites. Uniform and standardised parts, such as toilet fittings and door handles, would be fitted throughout many tower blocks; planners deemed that buying in bulk would reduce overall costs.

Another key aspect of the tower block vision was the Brutalist architectural method, popular with architects and planners at the time. The Brutalist emphasis led to the construction of stark and striking tower blocks with large sections of exposed concrete. Concrete was to be an integral part of the tower block designs; it could be poured on site, offering boundless flexibility to the building designers. To the planners, concrete was a silver bullet for the construction process – it was economical, and "was vaunted as being long-lasting, if not indestructible".

The Post-war British Tower Block Reality

Elmet Towers in Swarcliffe, Leeds, showing the dereliction that led to its demolition

Coleman's 1985 work argues that in trying to emulate Le Corbusier's ideas, the tower block planners only succeeded in encouraging social problems. Although architects and local authorities intended the opposite, tower blocks quickly became, as Hanley sharply stated, 'slums in the sky'. Due to demanding deadlines, complicated construction practices were rushed and many tower blocks experienced structural decay as a result – roofs leaked, concrete suffered spalling, steel corroded, and damp penetrated the buildings. Unfortunately, by replicating tower blocks across the nation, planners 'disastrously' replicated design faults. In many tower blocks, concrete quickly exhibited signs of decay; cracks soon formed and destabilised the buildings. The partial collapse of the Ronan Point tower block is an infamous example of the hasty and substandard construction that occurred in a number of the towers. The tower blocks quickly lost their "futuristic" look; concrete turned from the crisp white the designers had imagined to a dull grey, stained by pollution.

Poor design decisions ruined the anticipated benefits of the buildings. Open spaces, which were supposed to benefit the residents, were instead unattractive, unused and inadequately supervised. Residents felt it was difficult to maintain the large open spaces around the blocks because they realistically belonged to no one. Social problems increased as the tower blocks quickly degraded because of poor maintenance and an insecure communal environment. Apart from frequent break-downs, communal lifts were a source of fear for people travelling alone. It was a rarity to "enter a clean-smelling, undefaced lift". The tower blocks, many of which were on the periphery of the city, made residents feel isolated and cut off from society. Outsiders and newcomers were also affected; they felt the overbearing design of the tower blocks made them fearsome and unsociable.

Power argues that as a direct consequence of their design and construction, security problems were prevalent in many of the tower blocks. Break-ins, vandalism and muggings were common, which were aided by the buildings' concealed areas, the mazes of internal corridors, and dark corners. Police were often required in the tower blocks, but their infrequent presence did little to pacify towers rife with delinquency. To contain disruptive behaviour, local authorities began to place "problem families" in the same blocks; Hanley argues that this policy only led to "further alienation ... nihilism and a creeping sense of lawlessness". Dunleavy seconds this, suggesting that the mental health of long-term tower block residents may have been detrimentally affected.

While local authorities and their architects intended to create tower blocks that encouraged harmonious and vibrant communities, often the results were far from ideal. Post-war tower blocks were compromised from the outset by a combination of faults: local authorities advocated impractical architectural methods; design and construction faults were frequently reproduced; and there appeared to be a lack of understanding about the social consequences of certain design features. Collectively, these oversights transformed many tower blocks into undesirable places to live.

High-rise apartment buildings of the "Komplex Leipziger Straße" in Berlin-Mitte.

Fire Safety

Compliant protections for residents in tower blocks can be very expensive to retrofit, and lessors (such as local authorities) have been sued for lack of a current fire safety inspection, or implementation of its recommendations. Ed Davey MP commented regarding a life-claiming 2009 fire that many blocks in the UK remained inadequate. North Ayrshire Council has decided to remove plastic cladding from its buildings following a fatal fire in a tower block. Poor maintenance of electrical equipment, fire doors and other features has caused widespread fire hazards and residents are asked to pay part of the cost of putting this right even though the problem is due to poor maintenance and errors in design.

Tower blocks may be inherently more prone to casualties from a fire because people living on higher floors cannot escape fires easily and the fire brigade cannot reach the higher floors quickly. In buildings with more than a hundred residents, ensuring that every single resident acts responsibly to minimize fire risk is difficult; poorer residents in tower blocks may be tempted to use cheaper flammable fuels rather than electricity, they are also more likely to be smokers (carelessness with cigarettes is a major cause of home fires), and they are more likely to have old furniture, not made to modern fire safety standards. Fire safety legislation introduced in 2006 requires new high rise buildings to be built to higher safety standards with sprinkler systems; the same standards do not apply to pre-2006 tower blocks, which contain a greater proportion of poor people. Recent studies have investigated the combined use of egress components (e.g., stairs and elevators) to enhance the effectiveness of evacuation strategies in case of fire.

Towards the Present Day

The 31-story Red Road estate in Glasgow, the tallest public housing towers in Europe at the time of their construction. Like many 1960s British high-rises, they are now condemned.

In recent years, some council or ex-council high-rises in the United Kingdom, including Trellick Tower, Keeling House, Sivill House and The Barbican Estate, have become popular with young professionals due to their excellent views, desirable locations and architectural pedigrees, and now command high prices. There are plans to redevelop the Little London and Lovell Park areas on the fringes of Leeds city centre into luxury flats for 'Young Urban Professionals'. The plans entail demolishing all of the council housing and refurbishing the highrise flats. This demand has led to many councils rethinking plans regarding their demolition.

In Glasgow, similar initiatives have taken place – in 2011, the Glasgow Housing Association saved one of a cluster of three condemned tower blocks in the Ibrox area with the help of government grants to transform it into desirable rented housing for young professionals. Glasgow itself has taken a more measured approach to its high-rise housing stock, eschewing the mass demolition programmes being practiced in other British cities – favouring selective refurbishment of estates with secure futures and on buildings in good structural condition, only razing towers where absolutely necessary.

After a gap of around 30 years, new high-rise flats are once again being built in Belfast, Birmingham (some are for wealthy people, e.g. The Mailbox, others are student accommodation), Cardiff, Aberdeen, Glasgow, Leeds, Liverpool, London, Manchester, and Newcastle upon Tyne; but this time sometimes for wealthy professionals, rather than the "lower classes." Their developers market these properties by using the American term 'apartment buildings', perhaps in an effort to distance these newer buildings from the older tower blocks from the 1950s and 1960s. These are usually taller than their older counterparts and generally built in and around these provincial city centres. They are often glass and aluminium clad. Tonight with Trevor McDonald highlighted that in Leeds and Manchester (perhaps the cities that had seen most development) only approximately half were occupied and with owner occupation often being as low as 10%. In Southend there are controversial plans to build new tower blocks for social housing, a supporter claims this will be cheaper than paying to house poor people with private landlords while opponents fear the creation of Ghettos.

Ireland

Flats in Ballymun, Dublin, Ireland

Republic of Ireland

In Dublin, the Ballymun Flats were built between 1966 and 1969: seven 15-story towers, nineteen 8-story blocks and ten 4-story blocks. These were the "seven towers" referred to in the U2 song

"Running to Stand Still". They have since been demolished. Inner Dublin flat complexes include Sheriff Street (demolished), Fatima Mansions (demolished and redeveloped), St. Joseph's Gardens (demolished; replaced by Killarney Court flat complex), St. Teresa's Gardens, Dolphin House, Liberty House, St. Michael's Estate and O'Devaney Gardens and a lot more mainly throughout the North and South Inner City of Dublin. Suburban flat complexes were built exclusively on the northside of the city in Ballymun, Coolock and Kilbarrack. These flats were badly affected by a heroin epidemic that hit working-class areas of Dublin in the 1980s and early 90s.

Over the last five years the largest cities such as Dublin, Cork, Limerick and Galway have witnessed new large apartment building. Some large towns such as Navan, Drogheda, Dundalk and Mullingar have also witnessed lots of modern apartments being built.

Northern Ireland

Tower blocks in Northern Ireland were never built to the frequency as in cities on the island of Great Britain. Most tower blocks and flat complexes are found in Belfast although many of these have been demolished since the 1990s and replaced with traditional public housing units. The mid-rise Divis flats complex in west Belfast was built in between 1968 and 1972. It was demolished in the early 1990s as the residents demanded new houses due to mounting problems with the flats. Divis Tower, built separately in 1966, still stands, however; and in 2007 work began to convert the former British Army base at the top two floors into new dwellings. Divis Tower was for several decades Ireland's tallest residential building, having since being surpassed by the privately owned Obel Tower in the city centre. In the north of the City, the iconic 7 towers complex in the New Lodge remains, although so too the problems that residents face, such as poor piping and limited sanitation. Farther north, the four tower blocks in Rathcoole dominate the local skyline, while in south Belfast, the tower blocks in Seymour Hill, Belvoir & Finaghy remain standing.

Most of the aforementioned high rise flats in the city were built by the Northern Ireland Housing Trust (NIHT) as part of overspill housing schemes, the first such development being the pair of point blocks in East Belfast's Cregagh estate. These eleven-story towers were completed in 1961 and were the first tall council housing blocks on the island of Ireland. The NIHT also designed the inner city Divis Flats complex. The six to eight story deck access flats that comprised most of the Divis estate were of poor build quality and were all demolished by the early 1990s. Similar slab blocks were built by the NIHT in East Belfast (Tullycarnet) and Derry's Bogside area, all four of which have been demolished.

Belfast Corporation constructed seven tower blocks on the former Victoria Barracks site in the New Lodge district. While the Corporation built some mid-rise flats as part of slum clearance schemes (most notably the now demolished Unity Flats and the 'Weetabix Flats in the Shankill area), New Lodge would be their only high rise project in the inner city with three more in outlying areas of the city during the 1960s, two being in Mount Vernon in North Belfast and one being in the Clarawood estate, East Belfast. The Royal Hospital built three thirteen-story towers for use as staff accommodation, prominently located adjacent to the M2 Motorway at Broadway. Belfast City Hospital also constructed a high rise slab block which since privatisation has been named Bradbury Court, formerly known as Erskine House. Queens University Belfast built several eleven storey towers at their Queens Elms student accommodation. Of the three sixteen-story point blocks of Larne Borough Council in the late 1960s, only one remains.

Asia

Residential tower complexes are common in Asian countries such as China, India, Indonesia, Taiwan, Singapore, Japan, and South Korea as urban densities are very high. In Singapore and urban Hong Kong, land prices are so high that a large portion of the population lives in high-rise apartments. In fact, over 60% of Hong Kong residents live in apartments, many of them condominiums.

Sarah Williams Goldhagen (2012) celebrated the work of innovative architecture firms such as WOHA (based in Singapore), Mass Studies (based in Seoul), Amateur Architecture Studio (based in Hangzhou), and the New York City-based Steven Holl in the transformation of residential towers into "vertical communities" or "vertical cities in the sky" providing aesthetic, unusually designed silhouettes on the skyline, comfortable private spaces and attractive public spaces. None of these "functional, handsome, and humane high-rise residential buildings" are affordable housing.

South Korea

Typical Tower block apartment in South Korea

In South Korea the tower blocks are called Apartment Complex. The first residential towers began to be built after the Korean War. The South Korean government needed to build many apartment complexes in the cities to be able to accommodate the citizens. In the 60 years since, as the population increased considerably, tower blocks have become more common. This time however the new tower blocks integrated shopping malls, parking systems and other convenient facilities.

Samsung Tower Palace in Seoul, South Korea, is the tallest apartment complex in Asia.

In Seoul, approximately 80% of its residents live in apartment complexes which comprise 98% of recent residential construction. Seoul proper is noted for its population density, almost twice that of New York and eight times greater than Rome, though slightly less than Paris. Its metropolitan area is the densest in the OECD.

China

The 2012 Pritzker Prize was awarded to Chinese architect Wang Shu. Among his winning designs is the Vertical Courtyard Apartments, six 26-story towers built in Hangzhou, China, by his architectural firm, Amateur Architecture Studio. "These towers were designed to house two-storey apartments, in which every inhabitant would enjoy "the illusion of living on the second floor", accomplished by folding concrete floor planes (like "bamboo mats," claims the firm), so that every third story opens into a private courtyard. In the larger towers, the two-story units are stacked slightly askew, adding to the visual interest of the variegated façades (Goldhagen 2012)."

Americas

Residential buildings in Santa Fe, Mexico City, Mexico

Canada

In Canada large multi-family buildings are usually known as *apartment buildings* or *apartment blocks* if they are rented from one common landowner, or *condominiums* or *condo towers* if each dwelling unit is individually owned; they may be called *low-rise* (or *walk-up*), *mid-rise*, *high-rise*, or *skyscraper* depending on their height. Tall residential towers are a staple building type in all large cities. Their relative prominence in Canadian cities varies substantially, however. In general, more populated cities have more high-rises than smaller cities, due to a relative the scarcity of land and a greater demand for housing.

However, some cities such as Quebec City, and Halifax have fewer high-rise buildings due to several factors: a focus on historic preservation, height restrictions, and lower growth rates. In middle-sized cities with a relatively low population density, such as Calgary, Edmonton, Winnipeg, or Hamilton, there are more apartment towers but they are greatly outnumbered by single-family houses. Most of the largest residential towers in Canada are found in Montreal, Toronto, and Vancouver—the country's most densely populated cities.

Toronto contains the second largest concentration of high-rise apartment buildings in North America (after New York). In Canada, like in other New World countries, but unlike Western Europe, most high-rise towers are located in the city centre (or "downtown"), where smaller, older buildings were demolished to make way in redevelopment schemes.

United States

432 Park Avenue, the tallest residential high-rise tower in the world, under construction in July 2014

In the United States tower blocks are commonly referred to as “midrise” or “highrise apartment buildings”, depending on their height, while buildings that house fewer flats (apartments), or are not as tall as the tower blocks, are called “lowrise apartment buildings”. Specifically, “midrise” buildings are as tall as the streets are wide, allowing 5 hours of sunlight on the street.

Some of the first residential towers were the Castle Village towers in Manhattan, New York City, completed in 1939. Their cross-shaped design was copied in towers in Parkchester and Stuyvesant Town residential developments.

The government’s experiments in the 1960s and 70s to use high-rise apartments as a means of providing the housing solution for the poor broadly resulted in failure. Made in the tower in the park style, all but a few high-rise housing projects in the nation’s largest cities, such as Cabrini–Green and Robert Taylor Homes in Chicago, Penn South in New York City and the Desire projects in New Orleans, fell victim to the “ghettofication” and are now being torn down, renovated, or replaced. Another example is the former Pruitt–Igoe complex in St. Louis, torn down in the 1970s.

In contrast to their public housing cousins, commercially developed high-rise apartment buildings continue to flourish in cities around the country largely due to high land prices and the housing boom of the 2000s. The Upper East Side in New York City and Chicago’s Gold Coast, both featuring high-rise apartments, are the wealthiest urban neighborhoods in the United States.

Currently, the tallest residential building in the world is 432 Park Avenue located in New York City, United States of America, having a height of 1396 feet with the highest occupied floor at 1287 feet.

Australia

High-rise living in Australia was limited to small pockets of bohemian inner Sydney until the 1960s, where a short-lived fashion saw public housing tenants located in new high-rise developments, especially in Sydney and Melbourne. The buildings pictured along with four other 16-story blocks were constructed on behalf of the Royal Australian Navy and were available to sailors and their families for accommodation. Due to social problems within these blocks the Navy left and the Department of Housing took charge and flats were let to low income and immigrant families. During the 1980s many people escaping communism in Eastern block countries were housed in these buildings. Developers have enthusiastically adopted the term "apartment" for these new high-rise blocks, perhaps to avoid the stigma still attached to housing commission flats.

Housing commission towers in Waterloo, Sydney, Australia

Streets in the Sky

"Street in the sky" at Park Hill

Streets in the sky is a style of architecture that emerged in Britain in the 1960s and 1970s. Generally built to replace run-down terraced housing, the new designs included not only modern improvements such as inside toilets, but also shops and other community facilities within high-rise blocks. Examples of the buildings and developments are Trellick Tower, Balfron Tower, Robin Hood Gardens and Keeling House in London, Hunslet Grange in Leeds and Park Hill, Sheffield. These were an attempt to develop a new architecture, differentiated from earlier large housing estates, such as Quarry Hill flats in Leeds. Alison and Peter Smithson were the architects of Robin Hood Gardens. Another large example, the Aylesbury Estate in South London, built in 1970, is about to be demolished. The Hulme Crescents in Manchester were the largest social housing scheme in Europe when built in 1972 but lasted just 22 years. The Crescents are recognised as one of the worst social housing schemes in British history, marred by design and practical problems; they severely challenged the viability of the Streets in the Sky ideal.

Deck Access

Deck access is a type of flat that is accessed from a walkway that is open to the elements, as opposed to flats that are accessed from fully enclosed internal corridors. Deck access blocks of flats are usually fairly low-rise structures. The decks can vary from simple walkways, which may be covered or uncovered, to decks wide enough for small vehicles. The best known example of deck-access flats in the UK is Park Hill, Sheffield, where the decks are wide enough to allow electric vehicles; however, the design is inspired by French Modernist architect Le Corbusier, particularly his Unite D'Habitation in Marseilles.

Green Tower Blocks

Salford tower blocks in 2001. In the UK, tower blocks were mostly built between the 1950s and 70s.

Green tower blocks have some scheme of living plants or green roofs or solar panels on their roofs and/or incorporate other environmentally friendly design features.

Architectural Engineering

César Pelli's Ratner Athletic Center uses cables and masts as load-bearing devices.

Architectural engineering, also known as building engineering, is the application of engineering principles and technology to building design and construction. Definitions of an architectural engineer may refer to:

- An engineer in the structural, mechanical, electrical, construction or other engineering fields of building design and construction.
- A licensed engineering professional in parts of the United States.
- Architectural engineers are those who work with other engineers and architects for the designing and construction of buildings.

Engineering for Designer

Structural Engineering

Structural engineering involves the analysis and design of physical objects (buildings, bridges, equipment supports, towers and walls). Those concentrating on buildings are responsible for the structural performance of a large part of the built environment and are, sometimes, informally referred to as "building engineers". Structural engineers require expertise in strength of materials and in the seismic design of structures covered by earthquake engineering. Architectural Engineers sometimes structural as one aspect of their designs; the structural discipline when practiced as a specialty works closely with architects and other engineering specialists.

Mechanical, Electrical, and Plumbing (MEP)

MEP room in a building

Mechanical engineering and electrical engineering engineers are specialists, commonly referred to as (mechanical, electrical, and plumbing) when engaged in the building design fields. Also known as "building services engineering" in the United Kingdom, Canada, and Australia. Mechanical engineers often design and oversee the heating, ventilation and air conditioning (HVAC), plumbing, and rain gutter systems. Plumbing designers often include design specifications for simple active fire protection systems, but for more complicated projects, fire protection engineers are often separately retained. Electrical engineers are responsible for the building's power distribution, telecommunication, fire alarm, signalization, lightning protection and control systems, as well as lighting systems.

The Architectural Engineer (PE) in the United States

In many jurisdictions of the United States, the architectural engineer is a licensed engineering professional. Usually a graduate of an architectural engineering university program preparing students to perform whole-building design in competition with architect-engineer teams; or for practice in one of structural, mechanical or electrical fields of building design, but with an appreciation of integrated architectural requirements.

Formal architectural engineering education, following the engineering model of earlier disciplines, developed in the late 19th century, and became widespread in the United States by the mid-20th century. With the establishment of a specific "architectural engineering" NCEES Professional Engineering registration examination in the 1990s, and first offering in April 2003, architectural engineering became recognized as a distinct engineering discipline in the United States. Architectural engineers are not entitled to practice architecture unless they are also licensed as architects.

The Architect as Architectural Engineer

In some countries, the practice of architecture includes planning, designing and overseeing the building's construction, and architecture, as a profession providing architectural services, is referred to as "architectural engineering". In Japan, a "first-class architect" plays the dual role of architect and building engineer, although the services of a licensed "structural design first-class architect" are required for buildings over a certain scale.

In some languages, such as Korean and Arabic, "architect" is literally translated as "architectural engineer". In some countries, an "architectural engineer" (such as the *ingegnere edile* in Italy) is entitled to practice architecture and is often referred to as an architect. These individuals are often also structural engineers. In other countries, such as Germany, Austria, Iran, and most of the Arab countries, architecture graduates receive an engineering degree (*Dipl.-Ing. – Diplom-Ingenieur*).

In Spain, an "architect" has a technical university education and legal powers to carry out building structure and facility projects.

In Brazil, architects and engineers used to share the same accreditation process (CONFEA – Federal Council of Engineering, Architecture and Agronomy). Now the Brazilian architects and urbanists have their own accreditation process (CAU – Architecture and Urbanism Council). Besides traditional architecture design training, Brazilian architecture courses also offer complementary training in engineering disciplines such as structural, electrical, hydraulic and mechanical engineering. After graduation, architects focus in architectural planning, yet they can be responsible to the whole building, when it concerns to small buildings (except in electric wiring, where the architect autonomy is limited to systems up to 30kVA, and it has to be done by an Electrical Engineer), applied to buildings, urban environment, built cultural heritage, landscape planning, interiorscape planning and regional planning.

In Greece licensed architectural engineers are graduates from architecture faculties that belong to the Polytechnic University, obtaining an "Engineering Diploma". They graduate after 5 years of studies and are fully entitled architects once they become members of the Technical Chamber of Greece (TEE – Τεχνικό Επιμελητήριο Ελλάδος). The Engineering Diploma equals a master's degree in ECTS units (300) according to the Bologna Accords.

Education

The architectural, structural, mechanical and electrical engineering branches each have well established educational requirements that are usually fulfilled by completion of a university program.

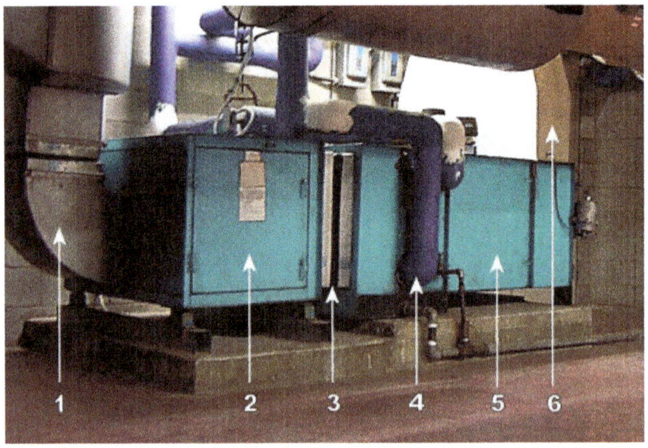

An air handling unit is used for the heating and cooling of air in a central location (click on image for legend). Bringing together knowledge of acoustic engineering and HVAC is one example of the multi-disciplined nature of architectural engineering

Architectural Engineering as a Single Integrated Field of Study

What differentiates architectural engineering from architecture (architect) as a separate and single, integrated field of study, compared to other engineering disciplines, is its multi-disciplinary engineering approach. Through training in and appreciation of architecture, the field seeks integration of building systems within its overall building design. Architectural engineering includes the design of building systems including heating, ventilation and air conditioning (HVAC), plumbing, fire protection, electrical, lighting, architectural acoustics, and structural systems. In some university programs, students are required to concentrate on one of the systems; in others, they can receive a generalist architectural or building engineering degree.

Structural Engineering

Structural engineering is mainly a sub-division of civil engineering in which structural engineers are trained to understand, predict, and calculate the stability, strength and rigidity of built structures for buildings and nonbuilding structures, to develop designs and integrate their design with that of other designers, and to supervise construction of projects on site. They can also be involved in the design of machinery, medical equipment, and vehicles where structural integrity affects functioning and safety.

Structural engineering theory is based upon applied physical laws and empirical knowledge of the structural performance of different materials and geometries. Structural engineering design utilizes a number of relatively simple structural elements to build complex structural systems. Structural engineers are responsible for making creative and efficient use of funds, structural elements and materials to achieve these goals.

The Eiffel Tower in Paris is a historical achievement of structural engineering.

Structural Engineer (Professional)

Structural engineers are responsible for engineering design and analysis. Entry-level structural engineers may design the individual structural elements of a structure, for example the beams, columns, and floors of a building. More experienced engineers may be responsible for the structural design and integrity of an entire system, such as a building.

Structural engineers often specialize in particular fields, such as bridge engineering, building engineering, pipeline engineering, industrial structures, or special mechanical structures such as vehicles, ships or aircraft.

Structural engineering has existed since humans first started to construct their own structures. It became a more defined and formalized profession with the emergence of the architecture as distinct profession from the engineering during the industrial revolution in the late 19th century. Until then, the architect and the structural engineer were usually one and the same thing - the master builder. Only with the development of specialized knowledge of structural theories that emerged during the 19th and early 20th centuries, did the professional structural engineers come into existence.

The role of a structural engineer today involves a significant understanding of both static and dynamic loading, and the structures that are available to resist them. The complexity of modern structures often requires a great deal of creativity from the engineer in order to ensure the structures support and resist the loads they are subjected to. A structural engineer will typically have a four or five year undergraduate degree, followed by a minimum of three years of professional practice before being considered fully qualified. Structural engineers are licensed or accredited by different learned societies and regulatory bodies around the world (for example, the Institution of Structural Engineers in the UK). Depending on the degree course they have studied and/or the jurisdiction they are seeking licensure in, they may be accredited (or licensed) as just structural engineers, or as civil engineers, or as both civil and structural engineers. Another international organisation is IABSE (International Association for Bridge and Structural Engineering). The aim of that association is to exchange knowledge and to advance the practice of structural engineering worldwide in the service of the profession and society.

History of Structural Engineering

Pont du Gard, France, a Roman era aqueduct circa 19 BC.

Structural engineering dates back to 2700 B.C.E. when the step pyramid for Pharaoh Djoser was built by Imhotep, the first engineer in history known by name. Pyramids were the most common major structures built by ancient civilizations because the structural form of a pyramid is inherently stable and can be almost infinitely scaled (as opposed to most other structural forms, which cannot be linearly increased in size in proportion to increased loads).

The structural stability of the pyramid is not primarily a result of its shape, as the integrity of the pyramid is intact as long as each of the stones is able to support the weight of the stone above it. The limestone blocks were taken from a quarry near the build site. Since the compressive strength of limestone is anywhere from 30 to 250 MPa (MPa = Pa * 10^6), the blocks will not fail under compression. Therefore, the structural strength of the pyramid stems from the material properties of the stones from which it was built rather than the pyramid's geometry.

Throughout ancient and medieval history most architectural design and construction was carried out by artisans, such as stone masons and carpenters, rising to the role of master builder. No theory of structures existed, and understanding of how structures stood up was extremely limited, and based almost entirely on empirical evidence of 'what had worked before'. Knowledge was retained by guilds and seldom supplanted by advances. Structures were repetitive, and increases in scale were incremental.

No record exists of the first calculations of the strength of structural members or the behavior of structural material, but the profession of structural engineer only really took shape with the Industrial Revolution and the re-invention of concrete. The physical sciences underlying structural engineering began to be understood in the Renaissance and have since developed into computer-based applications pioneered in the 1970s.

Timeline

- 1452–1519 Leonardo da Vinci made many contributions
- 1638: Galileo Galilei published the book *Two New Sciences* in which he examined the failure of simple structures

Galileo Galilei published the book *Two New Sciences* in which he examined the failure of simple structures

- 1660: Hooke's law by Robert Hooke
- 1687: Isaac Newton published *Philosophiae Naturalis Principia Mathematica* which contains the Newton's laws of motion

Isaac Newton published *Philosophiae Naturalis Principia Mathematica* which contains the Newton's laws of motion

- 1750: Euler–Bernoulli beam equation
- 1700–1782: Daniel Bernoulli introduced the principle of virtual work
- 1707–1783: Leonhard Euler developed the theory of buckling of columns

Leonhard Euler developed the theory of buckling of columns

- 1826: Claude-Louis Navier published a treatise on the elastic behaviors of structures
- 1873: Carlo Alberto Castigliano presented his dissertation "Intorno ai sistemi elastici", which contains his theorem for computing displacement as partial derivative of the strain energy. This theorem includes the method of "least work" as a special case
- 1874: Otto Mohr formalized the idea of a statically indeterminate structure.
- 1922: Timoshenko corrects the Euler-Bernoulli beam equation
- 1936: Hardy Cross' publication of the moment distribution method, an important innovation in the design of continuous frames.
- 1941: Alexander Hrennikoff solved the discretization of plane elasticity problems using a lattice framework
- 1942: R. Courant divided a domain into finite subregions
- 1956: J. Turner, R. W. Clough, H. C. Martin, and L. J. Topp's paper on the "Stiffness and Deflection of Complex Structures" introduces the name "finite-element method" and is widely recognized as the first comprehensive treatment of the method as it is known today

Structural Failure

The history of structural engineering contains many collapses and failures. Sometimes this is due to obvious negligence, as in the case of the Pétion-Ville school collapse, in which Rev. Fortin Augustin *"constructed the building all by himself, saying he didn't need an engineer as he had good knowledge of construction"* following a partial collapse of the three-story schoolhouse that sent neighbors fleeing. The final collapse killed 94 people, mostly children.

In other cases structural failures require careful study, and the results of these inquiries have resulted in improved practices and greater understanding of the science of structural engineering. Some such studies are the result of forensic engineering investigations where the original engineer seems to have done everything in accordance with the state of the profession and acceptable practice yet a failure still eventuated. A famous case of structural knowledge and practice being advanced in this manner can be found in a series of failures involving box girders which collapsed in Australia during the 1970s.

Specializations

Building Structures

Sydney Opera House, designed by Architect Jørn Utzon and structural design by Ove Arup & Partners

Structural building engineering includes all structural engineering related to the design of buildings. It is a branch of structural engineering closely affiliated with architecture.

Structural building engineering is primarily driven by the creative manipulation of materials and forms and the underlying mathematical and scientific ideas to achieve an end which fulfills its functional requirements and is structurally safe when subjected to all the loads it could reasonably be expected to experience. This is subtly different from architectural design, which is driven by the creative manipulation of materials and forms, mass, space, volume, texture and light to achieve an end which is aesthetic, functional and often artistic.

Millennium Dome in London, UK, by Richard Rogers and Buro Happold

The architect is usually the lead designer on buildings, with a structural engineer employed as a sub-consultant. The degree to which each discipline actually leads the design depends heavily on the type of structure. Many structures are structurally simple and led by architecture, such as multi-storey office buildings and housing, while other structures, such as tensile structures, shells and gridshells are heavily dependent on their form for their strength, and the engineer may have a more significant influence on the form, and hence much of the aesthetic, than the architect.

Burj Khalifa, in Dubai, the world's tallest building, shown under construction in 2007 (since completed)

The structural design for a building must ensure that the building is able to stand up safely, able to function without excessive deflections or movements which may cause fatigue of structural elements, cracking or failure of fixtures, fittings or partitions, or discomfort for occupants. It must account for movements and forces due to temperature, creep, cracking and imposed loads. It must also ensure that the design is practically buildable within acceptable manufacturing tolerances of the materials. It must allow the architecture to work, and the building services to fit within the building and function (air conditioning, ventilation, smoke extract, electrics, lighting etc.). The structural design of a modern building can be extremely complex, and often requires a large team to complete.

Structural engineering specialties for buildings include:

- Earthquake engineering
- Façade engineering
- Fire engineering
- Roof engineering
- Tower engineering
- Wind engineering

Earthquake Engineering Structures

Earthquake engineering structures are those engineered to withstand earthquakes.

The main objectives of earthquake engineering are to understand the interaction of structures with the shaking ground, foresee the consequences of possible earthquakes, and design and construct the structures to perform during an earthquake.

Earthquake-proof pyramid El Castillo, Chichen Itza

Earthquake-proof structures are not necessarily extremely strong like the El Castillo pyramid at Chichen Itza shown above. In fact, many structures considered strong may in fact be stiff, which can result in poor seismic performance.

One important tool of earthquake engineering is base isolation, which allows the base of a structure to move freely with the ground.

Civil Engineering Structures

Civil structural engineering includes all structural engineering related to the built environment. It includes:

• Bridges • Dams • Earthworks • Foundations • Offshore structures • Pipelines	• Power stations • Railways • Retaining structures and walls • Roads • Tunnels • Waterways • reservoir • Water and wastewater infrastructure

The structural engineer is the lead designer on these structures, and often the sole designer. In the design of structures such as these, structural safety is of paramount importance (in the UK, designs for dams, nuclear power stations and bridges must be signed off by a chartered engineer).

Civil engineering structures are often subjected to very extreme forces, such as large variations in temperature, dynamic loads such as waves or traffic, or high pressures from water or compressed gases. They are also often constructed in corrosive environments, such as at sea, in industrial facilities or below ground.

Mechanical Structures

Car

Principles of structural engineering are applied to variety of mechanical (moveable) structures. The design of static structures assumes they always have the same geometry (in fact, so-called static structures can move significantly, and structural engineering design must take this into account where necessary), but the design of moveable or moving structures must account for fatigue, variation in the method in which load is resisted and significant deflections of structures.

Motorbike

The forces which parts of a machine are subjected to can vary significantly, and can do so at a great rate. The forces which a boat or aircraft are subjected to vary enormously and will do so thousands of times over the structure's lifetime. The structural design must ensure that such structures are able to endure such loading for their entire design life without failing.

These works can require mechanical structural engineering:

- Boilers and pressure vessels
- Coachworks and carriages
- Cranes
- Elevators
- Escalators
- Marine vessels and hulls

Aerospace Structures

An Airbus A380, the world's largest passenger airliner

Aerospace structure types include launch vehicles, (Atlas, Delta, Titan), missiles (ALCM, Harpoon), Hypersonic vehicles (Space Shuttle), military aircraft (F-16, F-18) and commercial aircraft (Boeing 777, MD-11). Aerospace structures typically consist of thin plates with stiffeners for the external surfaces, bulkheads and frames to support the shape and fasteners such as welds, rivets, screws and bolts to hold the components together.

Design of missile needs in depth understanding of Structural Analysis

Nanoscale Structures

A nanostructure is an object of intermediate size between molecular and microscopic (micrometer-sized) structures. In describing nanostructures it is necessary to differentiate between the number of dimensions on the nanoscale. Nanotextured surfaces have one dimension on the nanoscale, i.e., only the thickness of the surface of an object is between 0.1 and 100 nm. Nanotubes have two dimensions on the nanoscale, i.e., the diameter of the tube is between 0.1 and 100 nm; its length could be much greater. Finally, spherical nanoparticles have three dimensions on the nanoscale, i.e., the particle is between 0.1 and 100 nm in each spatial dimension. The terms nanoparticles and ultrafine particles (UFP) often are used synonymously although UFP can reach into the micrometre range. The term 'nanostructure' is often used when referring to magnetic technology.

Structural Engineering for Medical Science

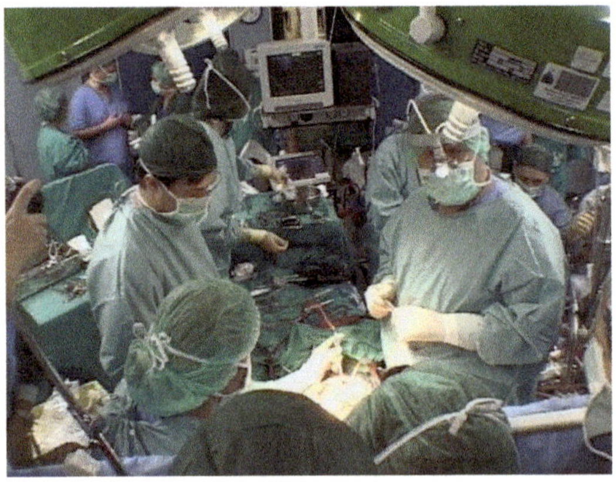

Designing Medical Equipment needs in-depth understanding of Structural Engineering

Medical equipment (also known as armamentarium) is designed to aid in the diagnosis, monitoring or treatment of medical conditions. There are several basic types: Diagnostic equipment includes medical imaging machines, used to aid in diagnosis ; equipment includes infusion pumps, medical lasers and LASIK surgical machines ; Medical monitors allow medical staff to measure a patient's medical state. Monitors may measure patient vital signs and other parameters including ECG, EEG, blood pressure, and dissolved gases in the blood ; Diagnostic Medical Equipment may also be used in the home for certain purposes, e.g. for the control of diabetes mellitus. A biomedical equipment technician (BMET) is a vital component of the healthcare delivery system. Employed primarily by hospitals, BMETs are the people responsible for maintaining a facility's medical equipment.

Structural Elements

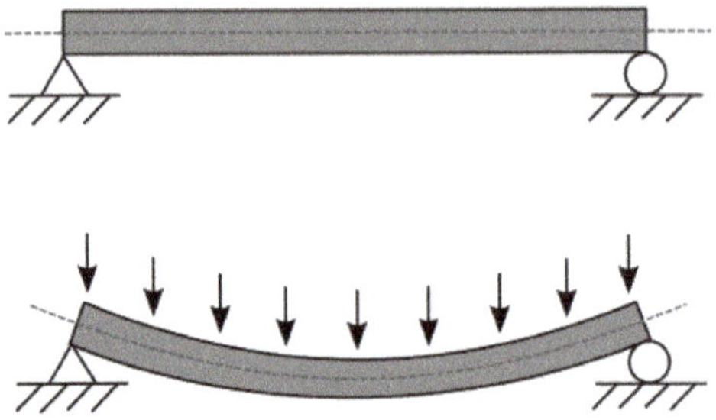

A statically determinate simply supported beam, bending under an evenly distributed load.

Any structure is essentially made up of only a small number of different types of elements:

- Columns
- Beams
- Plates
- Arches
- Shells
- Catenaries

Many of these elements can be classified according to form (straight, plane / curve) and dimensionality (one-dimensional / two-dimensional):

	One-dimensional		Two-dimensional	
	straight	curve	plane	curve
(predominantly) bending	beam	continuous arch	plate, concrete slab	lamina, dome
(predominant) tensile stress	rope, tie	Catenary	shell	
(predominant) compression	pier, column		Load-bearing wall	

Columns

Columns are elements that carry only axial force - compression - or both axial force and bending (which is technically called a beam-column but practically, just a column). The design of a column must check the axial capacity of the element, and the buckling capacity.

The buckling capacity is the capacity of the element to withstand the propensity to buckle. Its capacity depends upon its geometry, material, and the effective length of the column, which depends upon the restraint conditions at the top and bottom of the column. The effective length is where is the real length of the column and K is the factor dependent on the restraint conditions.

The capacity of a column to carry axial load depends on the degree of bending it is subjected to, and vice versa. This is represented on an interaction chart and is a complex non-linear relationship.

Beams

Little Belt: a truss bridge in Denmark

A beam may be defined as an element in which one dimension is much greater than the other two and the applied loads are usually normal to the main axis of the element. Beams and columns are called line elements and are often represented by simple lines in structural modeling.

- cantilevered (supported at one end only with a fixed connection)
- simply supported (supported vertically at each end; horizontally on only one to withstand friction, and able to rotate at the supports)
- fixed (supported at both ends by fixed connection; unable to rotate at the supports)

- continuous (supported by three or more supports)
- a combination of the above (ex. supported at one end and in the middle)

Beams are elements which carry pure bending only. Bending causes one part of the section of a beam (divided along its length) to go into compression and the other part into tension. The compression part must be designed to resist buckling and crushing, while the tension part must be able to adequately resist the tension.

Trusses

The McDonnell Planetarium by Gyo Obata in St Louis, Missouri, USA, a concrete shell structure

The 630 foot (192 m) high, stainless-clad (type 304) Gateway Arch in Saint Louis, Missouri

A truss is a structure comprising members and connection points or nodes. When members are connected at nodes and forces are applied at nodes members can act in tension or in compression. Members acting in compression are referred to as compression members or struts while members acting in tension are referred to as tension members or ties. Most trusses use gusset plates to connect intersecting elements. Gusset plates are relatively flexible and unable to transfer bending moments at the connections, thus allowing the truss members to act primarily in tension or compression.

Trusses are usually utilised in large-span structures, where it would be uneconomical to use solid beams.

Plates

Plates carry bending in two directions. A concrete flat slab is an example of a plate. Plates are understood by using continuum mechanics, but due to the complexity involved they are most often designed using a codified empirical approach, or computer analysis.

They can also be designed with yield line theory, where an assumed collapse mechanism is analysed to give an upper bound on the collapse load. This technique is used in practice but because the method provides an upper-bound, i.e. an unsafe prediction of the collapse load, for poorly conceived collapse mechanisms great care is needed to ensure that the assumed collapse mechanism is realistic.

Shells

Shells derive their strength from their form, and carry forces in compression in two directions. A dome is an example of a shell. They can be designed by making a hanging-chain model, which will act as a catenary in pure tension, and inverting the form to achieve pure compression.

Arches

Arches carry forces in compression in one direction only, which is why it is appropriate to build arches out of masonry. They are designed by ensuring that the line of thrust of the force remains within the depth of the arch. It is mainly used to increase the bountifulness of any structure.

Catenaries

Catenaries derive their strength from their form, and carry transverse forces in pure tension by deflecting (just as a tightrope will sag when someone walks on it). They are almost always cable or fabric structures. A fabric structure acts as a catenary in two directions.

Structural Engineering Theory

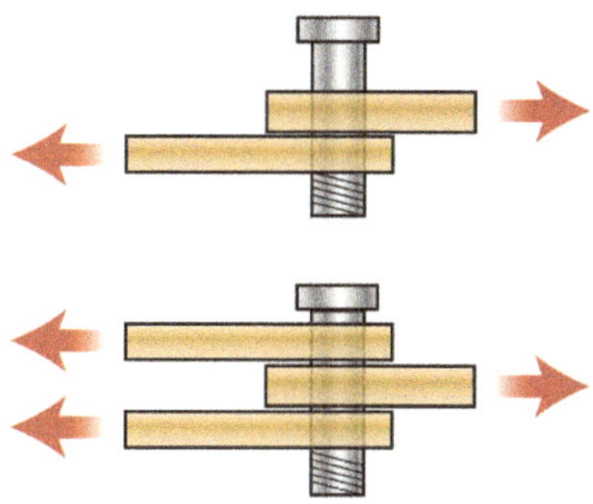

Figure of a bolt in shear stress. Top figure illustrates single shear, bottom figure illustrates double shear.

Structural engineering depends upon a detailed knowledge of applied mechanics, materials science and applied mathematics to understand and predict how structures support and resist self-weight and imposed loads. To apply the knowledge successfully a structural engineer generally requires detailed knowledge of relevant empirical and theoretical design codes, the techniques of structural analysis, as well as some knowledge of the corrosion resistance of the materials and structures, especially when those structures are exposed to the external environment. Since the 1990s, specialist software has become available to aid in the design of structures, with the functionality to assist in the drawing, analyzing and designing of structures with maximum precision; examples include AutoCAD, StaadPro, ETABS, Prokon, Revit Structure, Inducta RCB, etc. Such software may also take into consideration environmental loads, such as from earthquakes and winds.

Materials

Structural engineering depends on the knowledge of materials and their properties, in order to understand how different materials support and resist loads.

Common structural materials are:

- Iron: Wrought iron, Cast iron
- Concrete: Reinforced concrete, Prestressed concrete
- Alloy: Steel, Stainless steel
- Masonry
- Timber: Hardwood, Softwood
- Aluminium
- Composite materials: Plywood
- Other structural materials:Adobe, Bamboo, Carbon fibre, Fiber reinforced plastic, Mudbrick, Roofing materials

Heavy Engineering Construction

Heavy engineering construction includes primary infrastructural facilities such as transportation system railways, road, bridges, mass transit system, hydraulic structure (Dam, barrage, flood control system, hydro electric power, irrigation system) water treatment and water supply system, waste water treatment and its disposal, power line transmission.

Heavy engineering construction accounts for around 25-30 percent of total construction.

Construction is carried out by reputed big company.

Heavy engineering projects are mostly financed by government. These are primarily infrastuctural projects.

Design as well as construction in these projects is the primary job of civil engineers although inputs from other departments play an important role.

Heavy engineering project are highly equipment intensive and large quantities of materials are used.

Road

A road is a thoroughfare, route, or way on land between two places that has been paved or otherwise improved to allow travel by foot or some form of conveyance, including a motor vehicle, cart, bicycle, or horse.

Roads consist of one or two roadways (British English: carriageways), each with one or more lanes and any associated sidewalks (British English: pavement) and road verges.

Roads that are available for use by the public may be referred to as parkways, avenues, freeways, interstates, highways, or primary, secondary, and tertiary local roads.

Definitions

The Organisation for Economic Co-operation and Development (OECD) defines a road as "a line of communication (travelled way) using a stabilized base other than rails or air strips open to public traffic, primarily for the use of road motor vehicles running on their own wheels, "which includes" bridges, tunnels, supporting structures, junctions, crossings, interchanges, and toll roads, but not cycle paths."

In urban areas roads may diverge through a city or village and be named as streets, serving a dual function as urban space easement and route. Modern roads are normally smoothed, paved, or otherwise prepared to allow easy travel. Historically many roads were simply recognizable routes without any formal construction or maintenance.

United Kingdom

In the United Kingdom there is some ambiguity between the terms highway and road. The Highway code details rules for "road users". For the purposes of the English law, Highways Act 1980, which covers England and Wales but not Scotland or Northern Ireland, the term *road* is defined to be "any length of highway or of any other road to which the public has access, and includes bridges over which a road passes." This includes footpaths, bridleways and cycle tracks, and also road and driveways on private land and many car parks. Vehicle Excise Duty, a road use tax, is payable on some vehicles used on the public road.

The definition of a road depends on the definition of a highway; there is no formal definition for a highway in the relevant Act. A 1984 ruling said "the land over which a public right of way exists is known as a highway; and although most highways have been made up into roads, and most easements of way exist over footpaths, the presence or absence of a made road has nothing to do with the distinction. Another legal view is that while a highway historically included footpaths, bridleways, driftways, etc., it can now be used to mean those ways that allow the movement of motor-vehicles, and the term *rights of way* can be used to cover the wider usage.

United States

In the United States, laws distinguish between *public roads*, which are open to public use, and *private roads*, which are privately controlled.

History

Transfăgărăşan, named "the best road in the world"

The assertion that the first pathways were the trails made by animals has not been universally accepted; in many cases animals do not follow constant paths. Others believe that some roads originated from following animal trails. The Icknield Way is given as an example of this type of road origination, where man and animal both selected the same natural line. By about 10,000 BC, rough roads/pathways were used by human travelers.

A paved Roman road in Pompeii

- The world's oldest known paved road was constructed in Egypt some time between 2600 and 2200 BC.
- Stone-paved streets are found in the city of Ur in the Middle East dating back to 4000 BC.
- Corduroy roads (log roads) are found dating to 4000 BC in Glastonbury, England.
- The Sweet Track, a timber track causeway in England, is one of the oldest engineered roads discovered and the oldest timber trackway discovered in Northern Europe. Built in winter

3807 BC or spring 3806 BC, tree-ring dating (Dendrochronology) enabled very precise dating. It was claimed to be the oldest road in the world until the 2009 discovery of a 6,000-year-old trackway in Plumstead, London.

- Brick-paved streets were used in India as early as 3000 BC .
- In 500 BC, Darius I the Great started an extensive road system for Persia (Iran), including the Royal Road, which was one of the finest highways of its time. The road remained in use after Roman times.
- In ancient times, transport by river was far easier and faster than transport by road, especially considering the cost of road construction and the difference in carrying capacity between carts and river barges. A hybrid of road transport and ship transport beginning in about 1740 is the horse-drawn boat in which the horse follows a cleared path along the river bank.
- From about 312 BC, the Roman Empire built straight strong stone Roman roads throughout Europe and North Africa, in support of its military campaigns. At its peak the Roman Empire was connected by 29 major roads moving out from Rome and covering 78,000 kilometers or 52,964 Roman miles of paved roads.
- In the 8th century AD, many roads were built throughout the Arab Empire. The most sophisticated roads were those in Baghdad, which were paved with tar. Tar was derived from petroleum, accessed from oil fields in the region, through the chemical process of destructive distillation.
- The Highways Act 1555 in Britain transferred responsibility for maintaining roads from government to local parishes. This resulted in a poor and variable state of roads. To remedy this, the first of the "Turnpike trusts" was established around 1706, to build good roads and collect tolls from passing vehicles. Eventually there were approximately 1,100 trusts in Britain and some 36,800 km (22,870 miles) of engineered roads. The Rebecca Riots in Carmarthenshire and Rhayader from 1839 to 1844 contributed to a Royal Commission that led to the demise of the system in 1844, which coincided with the development of the UK railway system.

Design

The subject of road design is considered part of highway engineering. Structural road design is the science of designing a road for its environment in order to extend its longevity and reduce maintenance. The Shell pavement design method is used in many countries for the design of new asphalt roadsides.

Road Terminology

Alignment

the route of the road, defined as a series of horizontal tangents and curves.

Adverse camber

where a road slopes towards the outside of a bend, increasing the likelihood that vehicles travelling at speed will skid or topple. Usually only a temporary situation during road maintenance.

All-weather road

Unpaved road that is constructed of a material that does not create mud during rainfall.

Banked turn

Bicycle boulevard

A motorway open to all traffic but optimized for cycling

Cycling infrastructure

cycling-friendly infrastructure integrated into the roadway or in its own right of way

Camber (or crown)

The slope of the road surface downwards away from the centre of the road, so that surface water can flow freely to the edge of the carriageway, or on bends angling of the surface to lean traffic 'into the bend' reducing the chance of a skid.

Cant

Superelevation.

Cross slope

The slope of the pavement, expressed as units of rise per unit of run, or as a percentage.

Grade

longitudinal slope.

Belisha Beacon

An orange globe, lit at night, used to highlight a pedestrian crossing.

A beach road (Newcastle NSW Australia)

Bollard

Rigid posts that can be arranged in a line to close a road or path to vehicles above a certain width

Byway

Highway over which the public have a right to travel for vehicular and other kinds of traffic, but is used mainly as a footpath or bridleway

Bypass

Road that avoids or "bypasses" a built-up area, town, or village

Bottleneck

Section of a road with a carrying capacity substantially below that of other sections of the same road

Botts' dots

Non-reflective raised pavement marker used on roads

Cat's eye

reflective raised pavement marker used on roads

Chicane

Sequence of tight serpentine curves (usually an S-shape curve or a bus stop)

Chipseal

Road surface composed of a thin layer of crushed stone 'chips' and asphalt emulsion. It seals the surface and protects it from weather, but provides no structural strength. It is cheaper than asphalt concrete or concrete. In the United States it is usually only used on low volume rural roads

Corniche

Road on the side of a cliff or mountain, with the ground rising on one side and falling away on the other

Curb (kerb)

A raised edge at the side of the roadway.

Curb extension

(also kerb extension, bulb-out, nib, elephant ear, curb bulge and blister) Traffic calming measure, intended to slow the speed of traffic and increase driver awareness, particularly in built-up and residential neighborhoods.

Drainage gradient

Farm-to-market road

a state road or county road that connects rural or agricultural areas to market towns.

Fork

(literally "fork in the road") Type of intersection where a road splits

Green lane

(UK) Unsurfaced road, may be so infrequently used that vegetation colonises freely, hence 'green'. Many green lanes are ancient routes that have existed for millennia.

Guide rail

Prevents vehicles from veering off the road into oncoming traffic, crashing against solid objects or falling from a road. Also called a guard rail or traffic barrier.

Gutter

a drainage channel usually at the edge of the road or along a median.

Interstate Highway System (United States)

System of Interstate and Defense Highways

Layby (Pullout, pull-off)

A paved area beside a main road where cars can stop temporarily to let another car pass.

Loose chippings

the hazard of stone chippings that have come loose

Median

On dual carriageway roads, including controlled-access highways, divided highways and many limited-access roads, the central reservation (British English), median (North American English), median strip (North American English and Australian English), neutral ground [Louisiana English] or central nature strip (Australian English): Area that separates opposing lanes of traffic

Mountain pass

A relatively low level route through a range of mountains

Milestone

One of a series of numbered markers placed along a road, often at regular intervals, showing the distance to destinations.

National Highway

Road built and maintained by a national authority.

Pavement

The road regarded as a geoconstruction. In the UK the term is road surface and the pavement is a pedestrian walkway alongside the road.

Pedestrian crossing

Designated point on a road where road marking or other means helps pedestrians cross safely

Pelican crossing

(officially *Pelicon crossing*) (UK) a PEdestrian LIght CONtrolled crossing.

Private highway

Highway owned and operated for profit by private industry

Private road

Road owned and maintained by a private individual, organization, or company rather than by a government

Profile

the vertical alignment of a road, expressed as a series of grades, connected by parabolic curves.

Public space

Place where anyone has a right to come without being excluded because of economic or social conditions

Ranch road

U.S. road that connects rural and agricultural areas to market towns

Road number

Often assigned to identify a stretch of public roads— often dependent on the type of road, with numbers differentiating between interstates, motorways, arterial thoroughfares, etc.

Road-traffic safety

Process to reduce the harm (deaths, injuries, and property damage) that result from vehicle crashes on public roads

Roadworks

Part or all of the road is occupied for work or maintenance

Roughness

Deviations from a true planar pavement surface, which affects vehicle suspension deflection, dynamic loading, ride quality, surface drainage and winter operations. Roughness have wavelengths ranging from 500 mm up to some 40 m. The upper limit may be as high as 350 m when considering motion sickness aspects; motion sickness is generated by motion with down to 0.1 Hz frequency; in an ambulance car driving 35 m/s (126 km/h), waves with up to 350 m will excite motion sickness.

Roundabout

a road junction where typically three or more roads are joined by a circular section of road. Traffic 'on the roundabout' has priority over traffic on approach roads, unless indicated otherwise. In countries where traffic drives on the left the roundabout is travelled in a clockwise direction. Also known as an *island* in parts of the UK.

Shoulder (also hard shoulder)

A clear, level area to the side of the roadway available for stopping if needed.

State highway

Road numbered by the state, falling below numbered national highways (like U.S. Routes) in the hierarchy *or* a road maintained by the state, including nationally-numbered highways

Traffic

Pedestrians, ridden or herded animals, vehicles, bicycles, and other conveyances using any road for purposes of travel.

Texture (roads)

Deviations from a true planar pavement surface, which affects the interaction between road and tire. Microtexture have wavelengths below 0.5 mm, Macrotexture below 50 mm and Megatexture below 500 mm.

Traffic calming

Set of strategies used by urban planners and traffic engineers to slow down or reduce motor vehicle traffic, thereby improving safety for pedestrians and bicyclists and improving the environment for residents

Traffic island

(UK) a small raised area used to help define the traffic flow, which may also act as a refuge for pedestrians crossing the carriageway or a location for signs, barriers or lights—a synonym for roundabout in some parts of the UK

Traffic light

Also known as a traffic signal, stop light, stop-and-go lights—a signaling device at a road intersection, pedestrian crossing, or other location that assigns right of way to different approaches to an intersection

Zebra crossing (UK)

a pedestrian crossing marked by black and white stripes on the carriageway

Construction

Road construction requires the creation of an engineered continuous right-of-way or roadbed, overcoming geographic obstacles and having grades low enough to permit vehicle or foot travel. (pg15) and may be required to meet standards set by law or official guidelines. The process is often begun with the removal of earth and rock by digging or blasting, construction of embankments, bridges and tunnels, and removal of vegetation (this may involve deforestation) and followed by the laying of pavement material. A variety of road building equipment is employed in road building.

Surveyor at work with a leveling instrument

After design, approval, planning, legal and environmental considerations have been addressed alignment of the road is set out by a surveyor. The radii and gradient are designed and staked out to best suit the natural ground levels and minimize the amount of cut and fill. Great care is taken to preserve reference Benchmarks.

Asphalt layer and Hamm road roller

Roads are designed and built for primary use by vehicular and pedestrian traffic. Storm drainage and environmental considerations are a major concern. Erosion and sediment controls are constructed to prevent detrimental effects. Drainage lines are laid with sealed joints in the road easement with runoff coefficients and characteristics adequate for the land zoning and storm water system. Drainage systems must be capable of carrying the ultimate design flow from the upstream catchment with approval for the outfall from the appropriate authority to a watercourse, creek, river or the sea for drainage discharge.

Sub-base layer composed of cement-based material being applied during construction of the M8 motorway in Ireland

A borrow pit (source for obtaining fill, gravel, and rock) and a water source should be located near or in reasonable distance to the road construction site. Approval from local authorities may be required to draw water or for working (crushing and screening) of materials for construction needs. The top soil and vegetation is removed from the borrow pit and stockpiled for subsequent rehabilitation of the extraction area. Side slopes in the excavation area not steeper than one vertical to two horizontal for safety reasons.

Old road surfaces, fences, and buildings may need to be removed before construction can begin. Trees in the road construction area may be marked for retention. These protected trees should not have the topsoil within the area of the tree's drip line removed and the area should be kept clear of construction material and equipment. Compensation or replacement may be required if a protected tree is damaged. Much of the vegetation may be mulched and put aside for use during reinstatement. The topsoil is usually stripped and stockpiled nearby for rehabilitation of newly constructed embankments along the road. Stumps and roots are removed and holes filled as required before the earthwork begins. Final rehabilitation after road construction is completed will include seeding, planting, watering and other activities to reinstate the area to be consistent with the untouched surrounding areas.

Processes during earthwork include excavation, removal of material to spoil, filling, compacting, construction and trimming. If rock or other unsuitable material is discovered it is removed, moisture content is managed and replaced with standard fill compacted to meet the design requirements (generally 90-95% relative compaction). Blasting is not frequently used to excavate the road bed as the intact rock structure forms an ideal road base. When a depression must be filled to come up to the road grade the native bed is compacted after the topsoil has been removed. The fill is made by the "compacted layer method" where a layer of fill is spread then compacted to specifications, under saturated conditions. The process is repeated until the desired grade is reached.

General fill material should be free of organics, meet minimum California bearing ratio (CBR) results and have a low plasticity index. The lower fill generally comprises sand or a sand-rich mixture with fine gravel, which acts as an inhibitor to the growth of plants or other vegetable matter. The compacted fill also serves as lower-stratum drainage. Select second fill (sieved) should be composed of gravel, decomposed rock or broken rock below a specified particle size and be free of large lumps of clay. Sand clay fill may also be used. The road bed must be "proof rolled" after each layer of fill is compacted. If a roller passes over an area without creating visible deformation or spring the section is deemed to comply.

Typical pavement strata for a heavily traveled road

Geosynthetics such as geotextiles, geogrids and geocells are frequently used in the various pavement layers to improve road quality. These materials and methods are used in low-traffic private roadways as well as public roads and highways. Geosynthetics perform four main functions in roads: separation, reinforcement, filtration and drainage; which increase the pavement performance, reduce construction costs and decrease maintenance.

The completed road way is finished by paving or left with a gravel or other natural surface. The type of road surface is dependent on economic factors and expected usage. Safety improvements such as traffic signs, crash barriers, raised pavement markers and other forms of road surface marking are installed.

According to a May 2009 report by the American Association of State Highway and Transportation Officials (AASHTO) and TRIP—a national transportation research organization—driving on rough roads costs the average American motorist approximately $400 a year in extra vehicle operating costs. Drivers living in urban areas with populations more than 250,000 are paying upwards of $750 more annually because of accelerated vehicle deterioration, increased maintenance, additional fuel consumption, and tire wear caused by poor road conditions.

When a single carriageway road is converted into dual carriageway by building a second separate carriageway alongside the first, it is usually referred to as *duplication*, *twinning* or *doubling*. The original carriageway is changed from two-way to become one-way, while the new carriageway is one-way in the opposite direction. In the same way as converting railway lines from single track to double track, the new carriageway is not always constructed directly alongside the existing carriageway.

Maintenance

"Road works ahead" sign, typically used in Europe

Like all structures, roads deteriorate over time. Deterioration is primarily due to accumulated damage from vehicles, however environmental effects such as frost heaves, thermal cracking and oxidation often contribute. According to a series of experiments carried out in the late 1950s, called the AASHO Road Test, it was empirically determined that the effective damage done to the road is roughly proportional to the Fourth power of axle weight. A typical tractor-trailer weighing 80,000 pounds (36.287 t) with 8,000 pounds (3.629 t) on the steer axle and 36,000 pounds (16.329 t) on both of the tandem axle groups is expected to do 7,800 times more damage than a passenger vehicle with 2,000 pounds (0.907 t) on each axle. Potholes on roads are caused by rain damage and vehicle braking or related construction works.

Manual road repair taking place in Howrah, India

Pavements are designed for an expected service life or design life. In some parts of the United Kingdom the standard design life is 40 years for new bitumen and concrete pavement. Maintenance is considered in the whole life cost of the road with service at 10, 20 and 30 year milestones. Roads can be and are designed for a variety of lives (8-, 15-, 30-, and 60-year designs). When pavement lasts longer than its intended life, it may have been overbuilt, and the original costs may have been too high. When a pavement fails before its intended design life, the owner may have excessive repair and rehabilitation costs. Some asphalt pavements are designed as *perpetual pavements* with an expected structural life in excess of 50 years.

Line marking in rural India

Many asphalt pavements built over 35 years ago, despite not being specifically designed as a perpetual pavement, have remained in good condition long past their design life. Many concrete pavements built since the 1950s have significantly outlived their intended design lives. Some roads like Chicago, Illinois's "Wacker Drive", a major two-level viaduct in the downtown area, are being rebuilt with a designed service life of 100 years.

Virtually all roads require some form of maintenance before they come to the end of their service life. Pro-active agencies use pavement management techniques to continually monitor road conditions and schedule preventive maintenance treatments as needed to prolong the lifespan of their roads. Technically advanced agencies monitor the road network surface condition with sophisticated equipment such as laser/inertial Profilometers. These measurements include road curvature, cross slope, asperity, roughness, rutting and texture. This data is fed into a pavement management system, which recommends the best maintenance or construction treatment to correct the damage that has occurred.

Maintenance treatments for asphalt concrete generally include thin asphalt overlays, crack sealing, surface rejuvenating, fog sealing, micro milling or diamond grinding and surface treatments. Thin surfacing preserves, protects and improves the functional condition of the road while reducing the need for routing maintenance, leading to extended service life without increasing structural capacity.

Maintenance for the older concrete pavements that develop faults includes the technique called dowel bar retrofit. This involves cutting slots in the pavement at each joint, placing dowel bars in the slots, then filling them with concrete patching material. This method can extend the life of the concrete pavement for another 15 years.

Failure to maintain roads properly can create significant costs to society, in a 2009 report released by the American Association of State Highway and Transportation Officials (USA) about 50% of the roads in the USA are in bad condition with urban areas worse. The report estimates that urban drivers pay an average of $746/year on vehicle repairs while the average US motorist pays about $335/year. In contrast, the average motorist pays about $171/year in road maintenance taxes (based on 600 gallons/year and $0.285/gallon tax).

Slab Stabilization

Distress and serviceability loss on concrete roads can be caused by loss of support due to voids beneath the concrete pavement slabs. The voids usually occur near cracks or joints due to surface water infiltration. The most common causes of voids are pumping, consolidation, subgrade failure and bridge approach failure. Slab stabilization is a non-destructive method of solving this problem and is usually employed with other Concrete Pavement Restoration (CPR) methods including patching and diamond grinding. The technique restores support to concrete slabs by filing small voids that develop underneath the concrete slab at joints, cracks or the pavement edge.

The process consists of pumping a cementitous grout or polyurethane mixture through holes drilled through the slab. The grout can fill small voids beneath the slab and/or sub-base. The grout also displaces free water and helps keep water from saturating and weakening support under the joints and slab edge after stabilization is complete. The three steps for this method after finding the voids are locating and drilling holes, grout injection and post-testing the stabilized slabs.

Slab stabilization does not correct depressions, increase the design structural capacity, stop erosion or eliminate faulting. It does, however, restore the slab support, therefore, decreasing deflections under the load. Stabilization should only be performed at joints and cracks where loss of support exists. Visual inspection is the simplest manner to find voids. Signs that repair is needed are transverse joint faulting, corner breaks and shoulder drop off and lines at or near joints and cracks. Deflection testing is another common procedure utilized to locate voids. It is recommended to do this testing at night as during cooler temperatures, joints open, aggregate interlock diminishes and load deflections are at their highest.

Testing

Ground penetrating radar pulses electromagnetic waves into the pavement and measures and graphically displays the reflected signal. This can reveal voids and other defects.

The epoxy/core test, detects voids by visual and mechanical methods. It consists of drilling a 25 to 50 millimeter hole through the pavement into the sub-base with a dry-bit roto-hammer. Next, a two-part epoxy is poured into the hole—dyed for visual clarity. Once the epoxy hardens, technicians drill through the hole. If a void is present, the epoxy will stick to the core and provide physical evidence.

Common stabilization materials include pozzolan-cement grout and polyurethane. The requirements for slab stabilization are strength and the ability to flow into or expand to fill small voids. Colloidal mixing equipment is necessary to use the pozzolan-cement grouts. The contractor must place the grout using a positive-displacement injection pump or a non-pulsing progressive cavity pump. A drill is also necessary but it must produce a clean hole with no surface spalling or breakouts. The injection devices must include a grout packer capable of sealing the hole. The injection device must also have a return hose or a fast-control reverse switch, in case workers detect slab movement on the uplift gauge. The uplift beam helps to monitor the slab deflection and has to have sensitive dial gauges.

Joint Sealing

Also called joint and crack repair, this method's purpose is to minimize infiltration of surface water and incompressible material into the joint system. Joint sealants are also used to reduce dowel bar corrosion in Concrete Pavement Restoration (CPR) techniques. Successful resealing consists of old sealant removal, shaping and cleaning the reservoir, installing the backer rod and installing the sealant. Sawing, manual removal, plowing and cutting are methods used to remove the old sealant. Saws are used to shape the reservoir. When cleaning the reservoir, no dust, dirt or traces of old sealant should remain. Thus, it is recommended to water wash, sand-blast and then air blow to remove any sand, dirt or dust. The backer rod installation requires a double-wheeled, steel roller to insert the rod to the desired depth. After inserting the backer rod, the sealant is placed into the joint. There are various materials to choose for this method including hot pour bituminous liquid, silicone and preformed compression seals.

Safety Considerations

Pedestrian crossing, line markings and street furniture.

Careful design and construction of roads can increase road traffic safety and reduce the harm (deaths, injuries, and property damage) on the highway system from traffic collisions.

On neighborhood roads traffic calming, safety barriers, pedestrian crossings and cycle lanes can all protect pedestrians and cyclists.

Lane markers in some countries and states are marked with Cat's eyes or Botts dots, bright reflectors that do not fade like paint. Botts dots are not used where it is icy in the winter, because frost and snowplows can break the glue that holds them to the road, although they can be embedded in short, shallow trenches carved in the roadway, as is done in the mountainous regions of California.

For major roads risk can be reduced by providing limited access from properties and local roads, grade separated junctions and median dividers between opposite-direction traffic to reduce likelihood of head-on collisions.

The placement of energy attenuation devices (e.g. guardrails, wide grassy areas, sand barrels) is also common. Some road fixtures such as road signs and fire hydrants are designed to collapse on impact. Light poles are designed to break at the base rather than violently stop a car that hits them. Highway authorities may also remove larger trees from the immediate vicinity of the road. During heavy rains, if the elevation of the road surface isn't higher than the surrounding landscape, it may result in flooding.

Environmental Performance

Air pollution along Pasadena Highway in Los Angeles, California

Water management systems can be used to reduce the effect of pollutants from roads. Rainwater and snowmelt running off of roads tends to pick up gasoline, motor oil, heavy metals, trash and other pollutants and result in Water pollution. Road runoff is a major source of nickel, copper, zinc, cadmium, lead and polycyclic aromatic hydrocarbons (PAHs), which are created as combustion byproducts of gasoline and other fossil fuels.

A dual carriageway section of National Highway 8 connecting Delhi to Gurgaon

Careful design and construction of a road can reduce any negative environmental impacts.

Road after rain

De-icing chemicals and sand can run off into roadsides, contaminate groundwater and pollute surface waters; and road salts can be toxic to sensitive plants and animals. Sand applied to icy roads can be ground up by traffic into fine particulates and contribute to air pollution. Sand can alter stream bed environments, causing stress for the plants and animals that live there.

The A22(T) with line markings near Summer Hill, East Sussex, England

Roads are a chief source of environmental noise generation. In the early 1970s it was recognized that design of roads can be conducted to influence and minimize noise generation. Noise barriers are used to reduce noise pollution, in particular where roads are located close to built-up areas. Regulations can restrict the use of engine braking.

Road with guard rails in Kaluga Oblast, Russia

Motor vehicle emissions contribute air pollution. Concentrations of air pollutants and adverse respiratory health effects are greater near the road than at some distance away from the road. Road dust kicked up by vehicles may trigger allergic reactions. In addition, on-road transportation greenhouse gas emissions are the largest single cause of climate change, scientists say.

Road with traffic signs in the outskirts of Bern, Switzerland

NH 73 going Bangalore

Highway 401, a route with a collector / express setup

The wide Coastal Road in the Philippines now called the Manila-Cavite Expressway

Road in Jujuy Province, Argentina

Regulation

Right- and Left-hand Traffic

Traffic flows on the right or on the left side of the road depending on the country. In countries where traffic flows on the right, traffic signs are mostly on the right side of the road, roundabouts and traffic circles go counter-clockwise/anti-clockwise, and pedestrians crossing a two-way road should watch out for traffic from the left first. In countries where traffic flows on the left, the reverse is true.

About 33% of the world by population drive on the left, and 67% keep right. By road distances, about 28% drive on the left, and 72% on the right, even though originally most traffic drove on the left worldwide.

Economics

A city street in Mumbai, India with left-hand traffic

Transport economics is used to understand both the relationship between the transport system and the wider economy and the complex network effects when there are multiple paths and competing modes for both personal and freight (road/rail/air/ferry) and where Induced demand can result in increased on decreased transport levels when road provision is increased by building new roads or decreased (for example California State Route 480). Roads are generally built and maintained by the public sector using taxation although implementation may be through private contractors). or occasionally using road tolls.

Public-private partnerships are a way for communities to address the rising cost by injecting private funds into the infrastructure. There are four main ones:

- design/build
- design/build/operate/maintain
- design/build/finance/operate
- build/own/operate

Economics and society depend heavily on efficient roads. In the European Union (EU) 44% of all goods are moved by trucks over roads and 85% of all people are transported by cars, buses or coaches on roads. The term was also commonly used to refer to roadsteads, waterways that lent themselves to use by shipping.

Construction Costs

According to the New York State Thruway Authority, some sample per-mile costs to construct multi-lane roads in several US northeastern states were:

- Connecticut Turnpike — $3,449,000 per mile
- New Jersey Turnpike — $2,200,000 per mile
- Pennsylvania Turnpike (Delaware Extension) — $1,970,000 per mile
- Northern Indiana Toll Road — $1,790,000 per mile
- Garden State Parkway — $1,720,000 per mile
- Massachusetts Turnpike — $1,600,000 per mile
- Thruway, New York to Pennsylvania Line — $1,547,000 per mile
- Ohio Turnpike — $1,352,000 per mile
- Pennsylvania Turnpike (early construction) — $736,000 per mile

Statistics

The United States has the largest network of roads of any country with 4,050,717 miles (6,518,997 km) as of 2009. The Republic of India has the second largest road system in the world with 4,689,842 kilometres (2,914,133 mi) of road (2013). The People's Republic of China is third with 3,583,715 kilometres (2,226,817 mi) of road (2007). The Federative Republic of Brazil has the fourth largest road system in the world with 1,751,868 kilometres (1,088,560 mi) (2002). See List of countries by road network size. When looking only at expressways the National Trunk Highway System (NTHS) in China has a total length of 45,000 kilometres (28,000 mi) at the end of 2006, and 60,300 km at the end of 2008, second only to the United States with 90,000 kilometres (56,000 mi) in 2005.

Global Connectivity

Eurasia, Africa, North America, South America, and Australia each have an extensive road network that connects most cities. The North and South American road networks are separated by the Darién Gap, the only interruption in the Pan-American Highway. Eurasia and Africa are connected by roads on the Sinai Peninsula. The European Peninsula is connected to the Scandinavian Peninsula by the Øresund Bridge, and both have many connections to the mainland of Eurasia, including the bridges over the Bosphorus. Antarctica has very few roads and no continent-bridging network, though there are a few ice roads between bases, such as the South Pole Traverse. Bahrain is the only island country to be connected to a continental network by road (the King Fahd Causeway to Saudi Arabia). Even well-connected road networks are controlled by many different legal jurisdictions, and laws such as which side of the road to drive on vary accordingly.

Many populated domestic islands are connected to the mainland by bridges. A very long example is the 113-mile (181.9 km) Overseas Highway connecting many of the Florida Keys with the continental United States.

Even on mainlands, some settlements have no roads connecting with the primary continental network, due to natural obstacles like mountains or wetlands, remoteness, or general expense. Unpaved roads or lack of roads are more common in developing countries, and these can become impassible in wet conditions. As of 2014, only 43% of rural Africans have access to an all-season road. Due to steepness, mud, snow, or fords, roads can sometimes be passable only to four-wheel drive vehicles, those with snow chains or snow tires, or those capable of deep wading or amphibious operation.

Cities on the mainland of continents which do not have road access include:

- Iquitos, Peru, population 437,376 (2015) in the Amazon rainforest
- Juneau, Alaska, population 32,406 (2014)
- Nome, Alaska, population 3,788 (2014)
- Rankin Inlet, Nunavut, population 2,577 (2011)
- Supai, Arizona, population 208 (2010) in the Grand Canyon

Most disconnected settlements have local road networks connecting ports, buildings, and other points of interest.

Where demand for travel by road vehicle to a disconnected island or mainland settlement is high, roll-on/roll-off ferries are commonly available if the journey is relatively short. For long-distance trips, passengers usually travel by air and rent a car upon arrival. If facilities are available, vehicles and cargo can also be shipped to many disconnected settlements by boat, or air transport at much greater expense.. The island of Great Britain is connected to the European road network by Eurotunnel Shuttle - an example of a car shuttle train which is a service used in other parts of Europe to travel under mountains and over wetlands.

In polar areas, disconnected settlements are often more easily reached by snowmobile or dogsled in cold weather, which can produce sea ice that blocks ports, and bad weather that prevents flying. For example, resupply aircraft are only flown to Amundsen–Scott South Pole Station October to February, and many residents of coastal Alaska have bulk cargo shipped in only during the warmer months. Permanent darkness during the winter can also make long-distance travel more dangerous in polar areas. Continental road networks do reach into these areas, such as the Dalton Highway to the North Slope of Alaska, the R21 highway to Murmansk in Russia, and many roads in Scandinavia (though due to fjords water transport is sometimes faster). Large areas of Alaska, Canada, Greenland, and Siberia are sparsely connected. For example, all 25 communities of Nunavut are disconnected from each other and the main North American road network.

Road transport of people and cargo by may also be obstructed by border controls and travel restrictions. For example, travel from other parts of Asia to South Korea would require passage through the hostile country of North Korea. Moving between most countries in Africa and Eurasia would require passing through Egypt and Israel, which is a politically sensitive area.

Some places are intentionally car-free, and roads (if present) might be used by bicycles or pedestrians.

Roads are under construction to many remote places, such as the villages of the Annapurna Circuit, and a road was completed in 2013 to Mêdog County. Additional intercontinental and transoceanic fixed links have been proposed, including a Bering Strait crossing that would connect Eurasia-Africa and North America, a Malacca Strait Bridge to the largest island of Indonesia from Asia, and a Strait of Gibraltar crossing to connect Europe and Africa directly.

Transportation Engineering

The engineering of this roundabout in Bristol, England, attempts to make traffic flow free-moving

Transportation engineering or transport engineering is the application of technology and scientific principles to the planning, functional design, operation and management of facilities for any mode of transportation in order to provide for the safe, efficient, rapid, comfortable, convenient, economical, and environmentally compatible movement of people and goods (transport). It is a sub-discipline of civil engineering. The importance of transportation engineering within the civil engineering profession can be judged by the number of divisions in ASCE (American Society of Civil Engineers) that are directly related to transportation. There are six such divisions (Aerospace; Air Transportation; Highway; Pipeline; Waterway, Port, Coastal and Ocean; and Urban Transportation) representing one-third of the total 18 technical divisions within the ASCE (1987).

The planning aspects of transportation engineering relate to elements of urban planning, and involve technical forecasting decisions and political factors. Technical forecasting of passenger travel usually involves an urban transportation planning model, requiring the estimation of trip generation (how many trips for what purpose), trip distribution (destination choice, where is the traveler going), mode choice (what mode is being taken), and route assignment (which streets or routes are being used). More sophisticated forecasting can include other aspects of traveler decisions, including auto ownership, trip chaining (the decision to link individual trips together in a tour) and the choice of residential or business location (known as land use forecasting). Passenger trips are the focus of transportation engineering because they often represent the peak of demand on any transportation system.

A review of descriptions of the scope of various committees indicates that while facility planning and design continue to be the core of the transportation engineering field, such areas as operations planning, logistics, network analysis, financing, and policy analysis are also important to civil engineers, particularly to those working in highway and urban transportation. The National Council of Examiners for Engineering and Surveying (NCEES) list online the safety protocols, geometric design requirements, and signal timing.

Transportation engineering, as practiced by civil engineers, primarily involves planning, design, construction, maintenance, and operation of transportation facilities. The facilities support air, highway, railroad, pipeline, water, and even space transportation. The design aspects of transportation engineering include the sizing of transportation facilities (how many lanes or how much capacity the facility has), determining the materials and thickness used in pavement designing the geometry (vertical and horizontal alignment) of the roadway (or track).

Before any planning occurs the Engineer must take what is known as an inventory of the area or if it is appropriate, the previous system in place. This inventory or database must include information on (1)population, (2)land use, (3)economic activity, (4)transportation facilities and services, (5)travel patterns and volumes, (6)laws and ordinances, (7)regional financial resources, (8)community values and expectations. These inventories help the engineer create business models to complete accurate forecasts of the future conditions of the systemReview.

Operations and management involve traffic engineering, so that vehicles move smoothly on the road or track. Older techniques include signs, signals, markings, and tolling. Newer technologies involve intelligent transportation systems, including advanced traveler information systems (such as variable message signs), advanced traffic control systems (such as ramp meters), and vehicle infrastructure integration. Human factors are an aspect of transportation engineering, particularly concerning driver-vehicle interface and user interface of road signs, signals, and markings.

Highway Engineering

Engineers in this specialization:

- Handle the planning, design, construction, and operation of highways, roads, and other vehicular facilities as well as their related bicycle and pedestrian realms.
- Estimate the transportation needs of the public and then secure the funding for the project.
- Analyze locations of high traffic volumes and high collisions for safety and capacity.
- Use civil engineering principles to improve the transportation system.
- Utilizes the three design controls which are the drivers, the vehicles, and the roadways themselves.

Railroad Engineering

Railway engineers handle the design, construction, and operation of railroads and mass transit systems that use a fixed guideway (such as light rail or even monorails). Typical tasks would include determining horizontal and vertical alignment design, station location and design, and construction cost estimating. Railroad engineers can also move into the specialized field of train dispatching which focuses on train movement control.

Railway engineers also work to build a cleaner and safer transportation network by reinvesting and revitalizing the rail system to meet future demands. In the United States, railway engineers work with elected officials in Washington, D.C. on rail transportation issues to make sure that the rail system meets the country's transportation needs.

Port and Harbor Engineering

Port and harbor engineers handle the design, construction, and operation of ports, harbors, canals, and other maritime facilities. This is not to be confused with marine engineering.

Airport Engineering

Airport engineers design and construct airports. Airport engineers must account for the impacts and demands of aircraft in their design of airport facilities. These engineers must use the analysis of predominant wind direction to determine runway orientation, determine the size of runway border and safety areas, different wing tip to wing tip clearances for all gates and must designate the clear zones in the entire port.

Professional Societies

- Young Professionals in Transportation (YPT)
- Institute of Transportation Engineers (ITE)
- American Society for Highway Engineering (ASHE)
- American Society of Civil Engineers (ASCE)
- The American Association of State Highway and Transportation Officials (AASHTO)
- Eastern Asia Society for Transportation Studies (EASTS)

Earthquake Engineering

Earthquake engineering is an interdisciplinary branch of engineering that designs and analyzes structures, such as buildings and bridges, with earthquakes in mind. Its overall goal is to make such structures more resistant to earthquakes. An earthquake (or seismic) engineer aims to construct structures that will not be damaged in minor shaking and will avoid serious damage or collapse in a major earthquake. Earthquake engineering is the scientific field concerned with protecting society, the natural environment, and the man-made environment from earthquakes by limiting the seismic risk to socio-economically acceptable levels. Traditionally, it has been narrowly defined as the study of the behavior of structures and geo-structures subject to seismic loading; it is considered as a subset of structural engineering, geotechnical engineering, mechanical engineering, chemical engineering, applied physics, etc. However, the tremendous costs experienced in recent earthquakes have led to an expansion of its scope to encompass disciplines from the wider field of civil engineering, mechanical engineering and from the social sciences, especially sociology, political science, economics and finance.

The main objectives of earthquake engineering are:

- Foresee the potential consequences of strong earthquakes on urban areas and civil infrastructure.
- Design, construct and maintain structures to perform at earthquake exposure up to the expectations and in compliance with building codes.

A properly engineered structure does not necessarily have to be extremely strong or expensive. It has to be properly designed to withstand the seismic effects while sustaining an acceptable level of damage.

Shake-table crash testing of a regular building model (left) and a base-isolated building model (right) at UCSD

Seismic Loading

Tokyo Skytree, equipped with a tuned mass damper, is the world's tallest tower and is the world's second tallest structure.

Seismic loading means application of an earthquake-generated excitation on a structure (or geo-structure). It happens at contact surfaces of a structure either with the ground, with adjacent structures, or with gravity waves from tsunami. The loading that is expected at a given location on the Earth's surface is estimated by engineering seismology. It is related to the seismic hazard of the location.

Seismic Performance

Earthquake or seismic performance defines a structure's ability to sustain its main functions, such as its safety and serviceability, at and after a particular earthquake exposure. A structure is normally considered safe if it does not endanger the lives and well-being of those in or around it by partially or completely collapsing. A structure may be considered *serviceable* if it is able to fulfill its operational functions for which it was designed.

Basic concepts of the earthquake engineering, implemented in the major building codes, assume that a building should survive a rare, very severe earthquake by sustaining significant damage but without globally collapsing. On the other hand, it should remain operational for more frequent, but less severe seismic events.

Seismic Performance Assessment

Engineers need to know the quantified level of the actual or anticipated seismic performance associated with the direct damage to an individual building subject to a specified ground shaking. Such an assessment may be performed either experimentally or analytically.

Experimental Assessment

Experimental evaluations are expensive tests that are typically done by placing a (scaled) model of the structure on a shake-table that simulates the earth shaking and observing its behavior. Such kinds of experiments were first performed more than a century ago. Only recently has it become possible to perform 1:1 scale testing on full structures.

Due to the costly nature of such tests, they tend to be used mainly for understanding the seismic behavior of structures, validating models and verifying analysis methods. Thus, once properly validated, computational models and numerical procedures tend to carry the major burden for the seismic performance assessment of structures.

Analytical/Numerical Assessment

Snapshot from shake-table video of a 6-story non-ductile concrete building destructive testing

Seismic performance assessment or seismic structural analysis is a powerful tool of earthquake engineering which utilizes detailed modelling of the structure together with methods of structural analysis to gain a better understanding of seismic performance of building and non-building structures. The technique as a formal concept is a relatively recent development.

In general, seismic structural analysis is based on the methods of structural dynamics. For decades, the most prominent instrument of seismic analysis has been the earthquake response spectrum method which also contributed to the proposed building code's concept of today.

However, such methods are good only for linear elastic systems, being largely unable to model the structural behavior when damage (i.e., non-linearity) appears. Numerical *step-by-step integration* proved to be a more effective method of analysis for multi-degree-of-freedom structural systems with significant non-linearity under a transient process of ground motion excitation.

Basically, numerical analysis is conducted in order to evaluate the seismic performance of buildings. Performance evaluations are generally carried out by using nonlinear static pushover analysis or nonlinear time-history analysis. In such analyses, it is essential to achieve accurate non-linear modeling of structural components such as beams, columns, beam-column joints, shear walls etc. Thus, experimental results play an important role in determining the modeling parameters of individual components, especially those that are subject to significant non-linear deformations. The individual components are then assembled to create a full non-linear model of the structure. Thus created models are analyzed to evaluate the performance of buildings.

The capabilities of the structural analysis software are a major consideration in the above process as they restrict the possible component models, the analysis methods available and, most importantly, the numerical robustness. The latter becomes a major consideration for structures that venture into the non-linear range and approach global or local collapse as the numerical solution becomes increasingly unstable and thus difficult to reach. There are several commercially available Finite Element Analysis software's such as CSI-SAP2000 and CSI-PERFORM-3D and Scia Engineer-ECtools which can be used for the seismic performance evaluation of buildings. Moreover, there is research-based finite element analysis platforms such as OpenSees, RUAUMOKO and the older DRAIN-2D/3D, several of which are now open source.

Research for Earthquake Engineering

Shake-table testing of Friction Pendulum Bearings at EERC

Research for earthquake engineering means both field and analytical investigation or experimentation intended for discovery and scientific explanation of earthquake engineering related facts, revision of conventional concepts in the light of new findings, and practical application of the developed theories.

The National Science Foundation (NSF) is the main United States government agency that supports fundamental research and education in all fields of earthquake engineering. In particular, it focuses on experimental, analytical and computational research on design and performance enhancement of structural systems.

E-Defense Shake Table

The Earthquake Engineering Research Institute (EERI) is a leader in dissemination of earthquake engineering research related information both in the U.S. and globally.

A definitive list of earthquake engineering research related shaking tables around the world may be found in Experimental Facilities for Earthquake Engineering Simulation Worldwide. The most prominent of them is now E-Defense Shake Table in Japan.

Major U.S. Research Programs

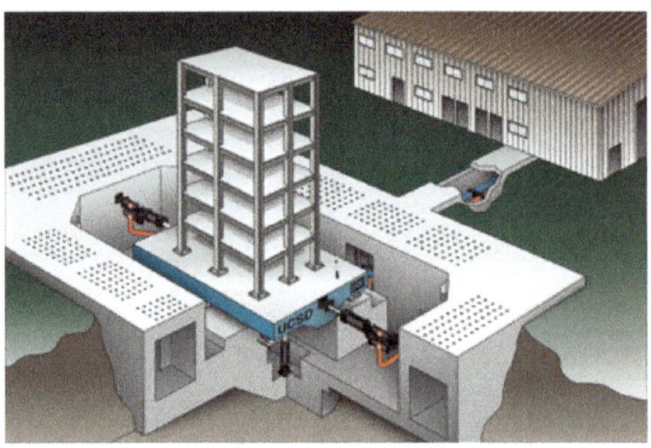

Large High Performance Outdoor Shake Table, UCSD, NEES network

NSF also supports the George E. Brown, Jr. Network for Earthquake Engineering Simulation

The NSF Hazard Mitigation and Structural Engineering program (HMSE) supports research on new technologies for improving the behavior and response of structural systems subject to earthquake hazards; fundamental research on safety and reliability of constructed systems; innovative developments in analysis and model based simulation of structural behavior and response including soil-structure interaction; design concepts that improve structure performance and flexibility; and application of new control techniques for structural systems.

(NEES) that advances knowledge discovery and innovation for earthquakes and tsunami loss reduction of the nation's civil infrastructure and new experimental simulation techniques and instrumentation.

The NEES network features 14 geographically-distributed, shared-use laboratories that support several types of experimental work: geotechnical centrifuge research, shake-table tests, large-scale structural testing, tsunami wave basin experiments, and field site research. Participating universities include: Cornell University; Lehigh University; Oregon State University; Rensselaer Polytechnic Institute; University at Buffalo, State University of New York; University of California, Berkeley; University of California, Davis; University of California, Los Angeles; University of California, San Diego; University of California, Santa Barbara; University of Illinois, Urbana-Champaign; University of Minnesota; University of Nevada, Reno; and the University of Texas, Austin.

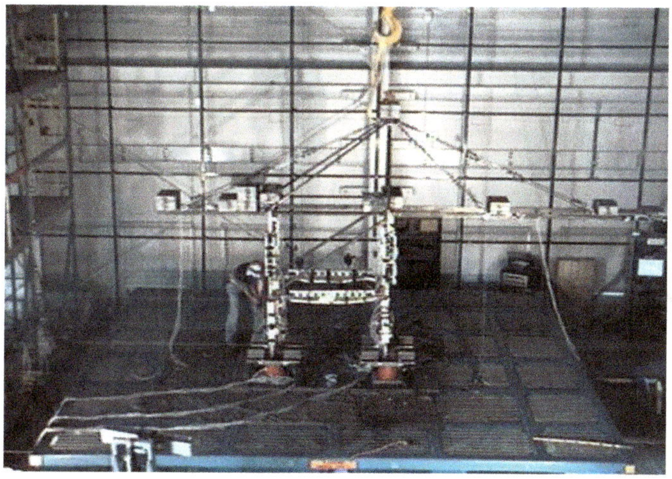

NEES at Buffalo testing facility

The equipment sites (labs) and a central data repository are connected to the global earthquake engineering community via the NEEShub website. The NEES website is powered by HUBzero software developed at Purdue University for nanoHUB specifically to help the scientific community share resources and collaborate. The cyberinfrastructure, connected via Internet2, provides interactive simulation tools, a simulation tool development area, a curated central data repository, animated presentations, user support, telepresence, mechanism for uploading and sharing resources, and statistics about users and usage patterns.

This cyberinfrastructure allows researchers to: securely store, organize and share data within a standardized framework in a central location; remotely observe and participate in experiments through the use of synchronized real-time data and video; collaborate with colleagues to facilitate the planning, performance, analysis, and publication of research experiments; and conduct computational and hybrid simulations that may combine the results of multiple distributed experiments and link physical experiments with computer simulations to enable the investigation of overall system performance.

These resources jointly provide the means for collaboration and discovery to improve the seismic design and performance of civil and mechanical infrastructure systems.

Earthquake Simulation

The very first earthquake simulations were performed by statically applying some horizontal inertia forces based on scaled peak ground accelerations to a mathematical model of a building. With the further development of computational technologies, static approaches began to give way to dynamic ones.

Dynamic experiments on building and non-building structures may be physical, like shake-table testing, or virtual ones. In both cases, to verify a structure's expected seismic performance, some researchers prefer to deal with so called "real time-histories" though the last cannot be "real" for a hypothetical earthquake specified by either a building code or by some particular research requirements. Therefore, there is a strong incentive to engage an earthquake simulation which is the seismic input that possesses only essential features of a real event.

Sometimes earthquake simulation is understood as a re-creation of local effects of a strong earth shaking.

Structure Simulation

Concurrent experiments with two building models which are *kinematically equivalent* to a real prototype.

Theoretical or experimental evaluation of anticipated seismic performance mostly requires a structure simulation which is based on the concept of structural likeness or similarity. Similarity is some degree of analogy or resemblance between two or more objects. The notion of similarity rests either on exact or approximate repetitions of patterns in the compared items.

In general, a building model is said to have similarity with the real object if the two share *geometric similarity*, *kinematic similarity* and *dynamic similarity*. The most vivid and effective type of similarity is the *kinematic* one. *Kinematic similarity* exists when the paths and velocities of moving particles of a model and its prototype are similar.

The ultimate level of *kinematic similarity* is *kinematic equivalence* when, in the case of earthquake engineering, time-histories of each story lateral displacements of the model and its prototype would be the same.

Seismic Vibration Control

Seismic vibration control is a set of technical means aimed to mitigate seismic impacts in building and non-building structures. All seismic vibration control devices may be classified as *passive*, *active* or *hybrid* where:

- *passive control devices* have no feedback capability between them, structural elements and the ground;
- *active control devices* incorporate real-time recording instrumentation on the ground integrated with earthquake input processing equipment and actuators within the structure;
- *hybrid control devices* have combined features of active and passive control systems.

When ground seismic waves reach up and start to penetrate a base of a building, their energy flow density, due to reflections, reduces dramatically: usually, up to 90%. However, the remaining portions of the incident waves during a major earthquake still bear a huge devastating potential.

After the seismic waves enter a superstructure, there are a number of ways to control them in order to soothe their damaging effect and improve the building's seismic performance, for instance:

- to dissipate the wave energy inside a superstructure with properly engineered dampers;
- to disperse the wave energy between a wider range of frequencies;
- to absorb the resonant portions of the whole wave frequencies band with the help of so-called *mass dampers*.

Mausoleum of Cyrus, the oldest base-isolated structure in the world

Devices of the last kind, abbreviated correspondingly as TMD for the tuned (*passive*), as AMD for the *active*, and as HMD for the *hybrid mass dampers*, have been studied and installed in high-rise buildings, predominantly in Japan, for a quarter of a century.

However, there is quite another approach: partial suppression of the seismic energy flow into the superstructure known as seismic or base isolation.

For this, some pads are inserted into or under all major load-carrying elements in the base of the building which should substantially decouple a superstructure from its substructure resting on a shaking ground.

The first evidence of earthquake protection by using the principle of base isolation was discovered in Pasargadae, a city in ancient Persia, now Iran, and dates back to the 6th century BCE. Below, there are some samples of seismic vibration control technologies of today.

Dry-stone Walls Control

Dry-stone walls of Machu Picchu Temple of the Sun, Peru

People of Inca civilization were masters of the polished 'dry-stone walls', called ashlar, where blocks of stone were cut to fit together tightly without any mortar. The Incas were among the best stonemasons the world has ever seen and many junctions in their masonry were so perfect that even blades of grass could not fit between the stones.

Peru is a highly seismic land and for centuries the mortar-free construction proved to be apparently more earthquake-resistant than using mortar. The stones of the dry-stone walls built by the Incas could move slightly and resettle without the walls collapsing, a passive structural control technique employing both the principle of energy dissipation and that of suppressing resonant amplifications.

Tuned Mass Damper

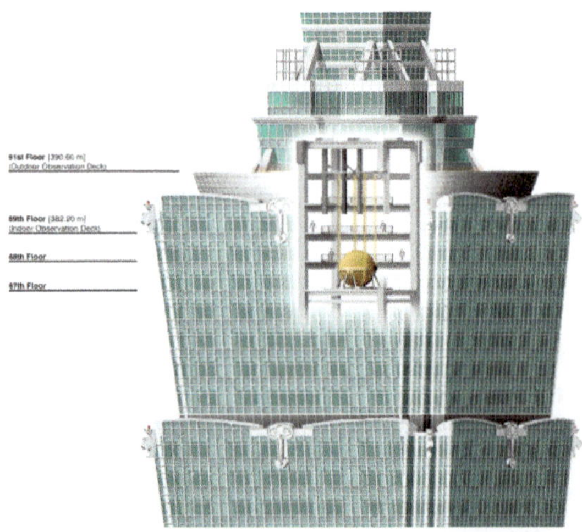

Tuned mass damper in Taipei 101, the world's third tallest skyscraper

Typically the tuned mass dampers are huge concrete blocks mounted in skyscrapers or other structures and moved in opposition to the resonance frequency oscillations of the structures by means of some sort of spring mechanism.

Taipei 101 skyscraper needs to withstand typhoon winds and earthquake tremors common in its area of the Asia-Pacific. For this purpose, a steel pendulum weighing 660 metric tones that serves as a tuned mass damper was designed and installed atop the structure. Suspended from the 92nd to the 88th floor, the pendulum sways to decrease resonant amplifications of lateral displacements in the building caused by earthquakes and strong gusts.

Hysteretic Dampers

Hysteretic damper is intended to provide better and more reliable seismic performance than that of a conventional structure at the expense of the seismic input energy dissipation. There are five major groups of hysteretic dampers used for the purpose, namely:

- Fluid viscous dampers (FVDs)

Viscous Dampers have the benefit of being a supplemental damping system. They have an oval hysteretic loop and the damping is velocity dependent. While some minor maintenance is potentially required, viscous dampers generally do not need to be replaced after an earthquake. While more expensive than other damping technologies they can be used for both seismic and wind loads and are the most commonly used hysteric damper.

- Friction dampers (FDs)

Friction dampers tend to be available in two major types, linear and rotational and dissipate energy by heat. The damper operates on the principle of a coulomb damper. Depending on the design, friction dampers can experience stick-slip phenomenon and Cold welding. The main disadvantage being that friction surfaces can wear over time and for this reason they are not recommended for dissipating wind loads. When used in seismic applications wear is not a problem and there is no required maintenance. They have a rectangular hysteretic loop and as long as the building is sufficiently elastic they tend to settle back to their original positions after an earthquake.

- Metallic yielding dampers (MYDs)

Metallic yielding dampers, as the name implies, yield in order to absorb the earthquake's energy. This type of damper absorbs a large amount of energy however they must be replaced after an earthquake and may prevent the building from settling back to its original position.

- Viscoelastic dampers (VEDs)

Viscoelastic dampers are useful in that they can be used for both wind and seismic applications, they are usually limited to small displacements. There is some concern as to the reliability of the technology as some brands have been banned from use in buildings in the United States.

- Straddlingpendulum dampers (swing)

Base Isolation

Base Isolation seeks to prevent the kinetic energy of the earthquake from being transferred into elastic energy in the building. These technologies do so by isolating the structure from the ground, thus enabling them to move somewhat independently. The degree to which the energy is transferred into the strucuture and how the energy is dissipated will vary depending on the technology used.

- Lead Rubber Bearing

LRB being tested at the UCSD Caltrans-SRMD facility

Lead Rubber Bearing or LRB is a type of base isolation employing a heavy damping. It was invented by Bill Robinson, a New Zealander.

Heavy damping mechanism incorporated in vibration control technologies and, particularly, in base isolation devices, is often considered a valuable source of suppressing vibrations thus enhancing a building's seismic performance. However, for the rather pliant systems such as base isolated structures, with a relatively low bearing stiffness but with a high damping, the so-called "damping force" may turn out the main pushing force at a strong earthquake. The video shows a Lead Rubber Bearing being tested at the UCSD Caltrans-SRMD facility. The bearing is made of rubber with a lead core. It was a uniaxial test in which the bearing was also under a full structure load. Many buildings and bridges, both in New Zealand and elsewhere, are protected with lead dampers and lead and rubber bearings. Te Papa Tongarewa, the national museum of New Zealand, and the New Zealand Parliament Buildings have been fitted with the bearings. Both are in Wellington which sits on an active fault.

- Springs-with-damper base isolator

Springs-with-damper base isolator installed under a three-story town-house, Santa Monica, California is shown on the photo taken prior to the 1994 Northridge earthquake exposure. It is a base isolation device conceptually similar to *Lead Rubber Bearing*.

One of two three-story town-houses like this, which was well instrumented for recording of both vertical and horizontal accelerations on its floors and the ground, has survived a severe shaking during the Northridge earthquake and left valuable recorded information for further study.

Springs-with-damper close-up

- Simple roller bearing

Simple roller bearing is a base isolation device which is intended for protection of various building and non-building structures against potentially damaging lateral impacts of strong earthquakes.

This metallic bearing support may be adapted, with certain precautions, as a seismic isolator to skyscrapers and buildings on soft ground. Recently, it has been employed under the name of *Metallic Roller Bearing* for a housing complex (17 stories) in Tokyo, Japan.

- Friction pendulum bearing

FPB shake-table testing

Friction Pendulum Bearing (FPB) is another name of Friction Pendulum System (FPS). It is based on three pillars:

- articulated friction slider;
- spherical concave sliding surface;
- enclosing cylinder for lateral displacement restraint.

Snapshot with the link to video clip of a shake-table testing of FPB system supporting a rigid building model is presented at the right.

Seismic Design

Seismic design is based on authorized engineering procedures, principles and criteria meant to design or retrofit structures subject to earthquake exposure. Those criteria are only consistent with the contemporary state of the knowledge about earthquake engineering structures. Therefore, a building design which exactly follows seismic code regulations does not guarantee safety against collapse or serious damage.

The price of poor seismic design may be enormous. Nevertheless, seismic design has always been a trial and error process whether it was based on physical laws or on empirical knowledge of the structural performance of different shapes and materials.

San Francisco City Hall destroyed by 1906 earthquake and fire.

San Francisco after the 1906 earthquake and fire

To practice seismic design, seismic analysis or seismic evaluation of new and existing civil engineering projects, an engineer should, normally, pass examination on *Seismic Principles* which, in the State of California, include:

- Seismic Data and Seismic Design Criteria
- Seismic Characteristics of Engineered Systems
- Seismic Forces

- Seismic Analysis Procedures
- Seismic Detailing and Construction Quality Control

To build up complex structural systems, seismic design largely uses the same relatively small number of basic structural elements (to say nothing of vibration control devices) as any non-seismic design project.

Normally, according to building codes, structures are designed to "withstand" the largest earthquake of a certain probability that is likely to occur at their location. This means the loss of life should be minimized by preventing collapse of the buildings.

Seismic design is carried out by understanding the possible failure modes of a structure and providing the structure with appropriate strength, stiffness, ductility, and configuration to ensure those modes cannot occur.

Seismic Design Requirements

Seismic design requirements depend on the type of the structure, locality of the project and its authorities which stipulate applicable seismic design codes and criteria. For instance, California Department of Transportation's requirements called *The Seismic Design Criteria* (SDC) and aimed at the design of new bridges in California incorporate an innovative seismic performance-based approach.

The Metsamor Nuclear Power Plant was closed after the 1988 Armenian earthquake

The most significant feature in the SDC design philosophy is a shift from a *force-based assessment* of seismic demand to a *displacement-based assessment* of demand and capacity. Thus, the newly adopted displacement approach is based on comparing the *elastic displacement* demand to the *inelastic displacement* capacity of the primary structural components while ensuring a minimum level of inelastic capacity at all potential plastic hinge locations.

In addition to the designed structure itself, seismic design requirements may include a *ground stabilization* underneath the structure: sometimes, heavily shaken ground breaks up which leads to collapse of the structure sitting upon it. The following topics should be of primary concerns: liquefaction; dynamic lateral earth pressures on retaining walls; seismic slope stability; earthquake-induced settlement.

Nuclear facilities should not jeopardise their safety in case of earthquakes or other hostile external events. Therefore, their seismic design is based on criteria far more stringent than those applying to non-nuclear facilities. The Fukushima I nuclear accidents and damage to other nuclear facilities that followed the 2011 Tōhoku earthquake and tsunami have, however, drawn attention to ongoing concerns over Japanese nuclear seismic design standards and caused other many governments to re-evaluate their nuclear programs. Doubt has also been expressed over the seismic evaluation and design of certain other plants, including the Fessenheim Nuclear Power Plant in France.

Failure Modes

Typical damage to unreinforced masonry buildings at earthquakes

Failure mode is the manner by which an earthquake induced failure is observed. It, generally, describes the way the failure occurs. Though costly and time consuming, learning from each real earthquake failure remains a routine recipe for advancement in *seismic design* methods. Below, some typical modes of earthquake-generated failures are presented.

Soft story collapse due to inadequate shear strength at ground level, Loma Prieta earthquake

The lack of reinforcement coupled with poor mortar and inadequate roof-to-wall ties can result in substantial damage to an unreinforced masonry building. Severely cracked or leaning walls are some of the most common earthquake damage. Also hazardous is the damage that may occur between the walls and roof or floor diaphragms. Separation between the framing and the walls can jeopardize the vertical support of roof and floor systems.

Soft story effect. Absence of adequate shear walls on the ground level caused damage to this structure. A close examination of the image reveals that the rough board siding, once covered by a brick veneer, has been completely dismantled from the studwall. Only the rigidity of the floor above combined with the support on the two hidden sides by continuous walls, not penetrated with large doors as on the street sides, is preventing full collapse of the structure.

Effects of soil liquefaction during the 1964 Niigata earthquake

Soil liquefaction. In the cases where the soil consists of loose granular deposited materials with the tendency to develop excessive hydrostatic pore water pressure of sufficient magnitude and compact, liquefaction of those loose saturated deposits may result in non-uniform settlements and tilting of structures. This caused major damage to thousands of buildings in Niigata, Japan during the 1964 earthquake.

Car smashed by landslide rock, 2008 Sichuan earthquake

Landslide rock fall. A landslide is a geological phenomenon which includes a wide range of ground movement, including rock falls. Typically, the action of gravity is the primary driving force for a landslide to occur though in this case there was another contributing factor which affected the original slope stability: the landslide required an *earthquake trigger* before being released.

Effects of pounding against adjacent building, Loma Prieta

Pounding against adjacent building. This is a photograph of the collapsed five-story tower, St. Joseph's Seminary, Los Altos, California which resulted in one fatality. During Loma Prieta earthquake, the tower pounded against the independently vibrating adjacent building behind. A possibility of pounding depends on both buildings' lateral displacements which should be accurately estimated and accounted for.

Effects of completely shattered joints of concrete frame, Northridge

At Northridge earthquake, the Kaiser Permanente concrete frame office building had joints completely shattered, revealing inadequate confinement steel, which resulted in the second story collapse. In the transverse direction, composite end shear walls, consisting of two wythes of brick and a layer of shotcrete that carried the lateral load, peeled apart because of inadequate through-ties and failed.

- Improper construction site on a foothill.
- Poor detailing of the reinforcement (lack of concrete confinement in the columns and at the beam-column joints, inadequate splice length).
- Seismically weak soft story at the first floor.
- Long cantilevers with heavy dead load.

shifting from foundation, Whittier

Sliding off foundations effect of a relatively rigid residential building structure during 1987 Whittier Narrows earthquake. The magnitude 5.9 earthquake pounded the Garvey West Apartment building in Monterey Park, California and shifted its superstructure about 10 inches to the east on its foundation.

Earthquake damage in Pichilemu.

If a superstructure is not mounted on a base isolation system, its shifting on the basement should be prevented.

Insufficient shear reinforcement let main rebars to buckle, Northridge

Reinforced concrete column burst at Northridge earthquake due to insufficient shear reinforcement mode which allows main reinforcement to buckle outwards. The deck unseated at the hinge and failed in shear. As a result, the La Cienega-Venice underpass section of the 10 Freeway collapsed.

Support-columns and upper deck failure, Loma Prieta earthquake

Loma Prieta earthquake: side view of reinforced concrete support-columns failure which triggered the upper deck collapse onto the lower deck of the two-level Cypress viaduct of Interstate Highway 880, Oakland, CA.

Failure of retaining wall due to ground movement, Loma Prieta

Retaining wall failure at Loma Prieta earthquake in Santa Cruz Mountains area: prominent north-west-trending extensional cracks up to 12 cm (4.7 in) wide in the concrete spillway to Austrian Dam, the north abutment.

Lateral spreading mode of ground failure, Loma Prieta

Ground shaking triggered soil liquefaction in a subsurface layer of sand, producing differential lateral and vertical movement in an overlying carapace of unliquified sand and silt. This mode of ground failure, termed lateral spreading, is a principal cause of liquefaction-related earthquake damage.

Beams and pier columns diagonal cracking, 2008 Sichuan earthquake

Severely damaged building of Agriculture Development Bank of China after 2008 Sichuan earthquake: most of the beams and pier columns are sheared. Large diagonal cracks in masonry and veneer are due to in-plane loads while abrupt settlement of the right end of the building should be attributed to a landfill which may be hazardous even without any earthquake.

Tsunami strikes Ao Nang,

Twofold tsunami impact: sea waves hydraulic pressure and inundation. Thus, the Indian Ocean earthquake of December 26, 2004, with the epicenter off the west coast of Sumatra, Indonesia, triggered a series of devastating tsunamis, killing more than 230,000 people in eleven countries by inundating surrounding coastal communities with huge waves up to 30 meters (100 feet) high.

Earthquake-resistant Construction

Earthquake construction means implementation of seismic design to enable building and non-building structures to live through the anticipated earthquake exposure up to the expectations and in compliance with the applicable building codes.

Construction of Pearl River Tower X-bracing to resist lateral forces of earthquakes and winds

Design and construction are intimately related. To achieve a good workmanship, detailing of the members and their connections should be as simple as possible. As any construction in general, earthquake construction is a process that consists of the building, retrofitting or assembling of infrastructure given the construction materials available.

The destabilizing action of an earthquake on constructions may be *direct* (seismic motion of the ground) or *indirect* (earthquake-induced landslides, soil liquefaction and waves of tsunami).

A structure might have all the appearances of stability, yet offer nothing but danger when an earthquake occurs. The crucial fact is that, for safety, earthquake-resistant construction techniques are as important as quality control and using correct materials. *Earthquake contractor* should be registered in the state/province/country of the project location (depending on local regulations), bonded and insured.

To minimize possible losses, construction process should be organized with keeping in mind that earthquake may strike any time prior to the end of construction.

Each construction project requires a qualified team of professionals who understand the basic features of seismic performance of different structures as well as construction management.

Adobe Structures

Around thirty percent of the world's population lives or works in earth-made construction. Adobe type of mud bricks is one of the oldest and most widely used building materials. The use of adobe is very common in some of the world's most hazard-prone regions, traditionally across Latin America, Africa, Indian subcontinent and other parts of Asia, Middle East and Southern Europe.

Adobe buildings are considered very vulnerable at strong quakes. However, multiple ways of seismic strengthening of new and existing adobe buildings are available.

Partially collapsed adobe building in Westmorland, California

Key factors for the improved seismic performance of adobe construction are:

- Quality of construction.
- Compact, box-type layout.
- Seismic reinforcement.

Limestone and Sandstone Structures

Base-isolated City and County Building, Salt Lake City, Utah

Limestone is very common in architecture, especially in North America and Europe. Many landmarks across the world are made of limestone. Many medieval churches and castles in Europe are made of limestone and sandstone masonry. They are the long-lasting materials but their rather heavy weight is not beneficial for adequate seismic performance.

Application of modern technology to seismic retrofitting can enhance the survivability of unreinforced masonry structures. As an example, from 1973 to 1989, the Salt Lake City and County Building in Utah was exhaustively renovated and repaired with an emphasis on preserving historical accuracy in appearance. This was done in concert with a seismic upgrade that placed the weak sandstone structure on base isolation foundation to better protect it from earthquake damage.

Timber Frame Structures

Anne Hvide's House, Denmark (1560)

Timber framing dates back thousands of years, and has been used in many parts of the world during various periods such as ancient Japan, Europe and medieval England in localities where timber was in good supply and building stone and the skills to work it were not.

The use of timber framing in buildings provides their complete skeletal framing which offers some structural benefits as the timber frame, if properly engineered, lends itself to better *seismic survivability*.

Light-frame Structures

A two-story wooden-frame for a residential building structure

Light-frame structures usually gain seismic resistance from rigid plywood shear walls and wood structural panel diaphragms. Special provisions for seismic load-resisting systems for all engineered wood structures requires consideration of diaphragm ratios, horizontal and vertical diaphragm shears, and connector/fastener values. In addition, collectors, or drag struts, to distribute shear along a diaphragm length are required.

Reinforced Masonry Structures

A construction system where steel reinforcement is embedded in the mortar joints of masonry or placed in holes and after filled with concrete or grout is called reinforced masonry.

Reinforced hollow masonry wall

The devastating 1933 Long Beach earthquake revealed that masonry construction should be improved immediately. Then, the California State Code made the reinforced masonry mandatory.

There are various practices and techniques to achieve reinforced masonry. The most common type is the reinforced hollow unit masonry. The effectiveness of both vertical and horizontal reinforcement strongly depends on the type and quality of the masonry, i.e. masonry units and mortar.

To achieve a ductile behavior of masonry, it is necessary that the shear strength of the wall is greater than the flexural strength.

Reinforced Concrete Structures

Stressed Ribbon pedestrian bridge over the Rogue River, Grants Pass, Oregon

Reinforced concrete is concrete in which steel reinforcement bars (rebars) or fibers have been incorporated to strengthen a material that would otherwise be brittle. It can be used to produce beams, columns, floors or bridges.

Prestressed concrete cable-stayed bridge over Yangtze river

Prestressed concrete is a kind of reinforced concrete used for overcoming concrete's natural weakness in tension. It can be applied to beams, floors or bridges with a longer span than is practical with ordinary reinforced concrete. Prestressing tendons (generally of high tensile steel cable or rods) are used to provide a clamping load which produces a compressive stress that offsets the tensile stress that the concrete compression member would, otherwise, experience due to a bending load.

To prevent catastrophic collapse in response earth shaking (in the interest of life safety), a traditional reinforced concrete frame should have ductile joints. Depending upon the methods used and the imposed seismic forces, such buildings may be immediately usable, require extensive repair, or may have to be demolished.

Prestressed Structures

Prestressed structure is the one whose overall integrity, stability and security depend, primarily, on a *prestressing*. *Prestressing* means the intentional creation of permanent stresses in a structure for the purpose of improving its performance under various service conditions.

Naturally pre-compressed exterior wall of Colosseum, Rome

There are the following basic types of prestressing:

- Pre-compression (mostly, with the own weight of a structure)

- Pretensioning with high-strength embedded tendons
- Post-tensioning with high-strength bonded or unbonded tendons

Today, the concept of prestressed structure is widely engaged in design of buildings, underground structures, TV towers, power stations, floating storage and offshore facilities, nuclear reactor vessels, and numerous kinds of bridge systems.

A beneficial idea of *prestressing* was, apparently, familiar to the ancient Rome architects; look, e.g., at the tall attic wall of Colosseum working as a stabilizing device for the wall piers beneath.

Steel Structures

Collapsed section of the San Francisco–Oakland Bay Bridge in response to Loma Prieta earthquake

Steel structures are considered mostly earthquake resistant but some failures have occurred. A great number of welded steel moment-resisting frame buildings, which looked earthquake-proof, surprisingly experienced brittle behavior and were hazardously damaged in the 1994 Northridge earthquake. After that, the Federal Emergency Management Agency (FEMA) initiated development of repair techniques and new design approaches to minimize damage to steel moment frame buildings in future earthquakes.

For structural steel seismic design based on Load and Resistance Factor Design (LRFD) approach, it is very important to assess ability of a structure to develop and maintain its bearing resistance in the inelastic range. A measure of this ability is ductility, which may be observed in a *material itself*, in a *structural element*, or to a *whole structure*.

As a consequence of Northridge earthquake experience, the American Institute of Steel Construction has introduced AISC 358 "Pre-Qualified Connections for Special and intermediate Steel Moment Frames." The AISC Seismic Design Provisions require that all Steel Moment Resisting Frames employ either connections contained in AISC 358, or the use of connections that have been subjected to pre-qualifying cyclic testing.

Prediction of Earthquake Losses

Earthquake loss estimation is usually defined as a Damage Ratio (DR) which is a ratio of the earthquake damage repair cost to the total value of a building. Probable Maximum Loss (PML) is a common term used for earthquake loss estimation, but it lacks a precise definition. In 1999, ASTM E2026 'Standard Guide for the Estimation of Building Damageability in Earthquakes' was produced in order to standardize the nomenclature for seismic loss estimation, as well as establish guidelines as to the review process and qualifications of the reviewer.

Earthquake loss estimations are also referred to as *Seismic Risk Assessments*. The risk assessment process generally involves determining the probability of various ground motions coupled with the vulnerability or damage of the building under those ground motions. The results are defined as a percent of building replacement value.

Industrial Project

Industrial Projects are highly specialized projects and it is dominated by large construction firms.

These projects include power plants(fossile fuel, nuclear power), petroleum refineries and petro-chemical plants, fertilizer plants, large steel plants, heavy manufacturing plants, aircraft industry.

Industrial projects accounts for only 10 percent of the total construction projects, but it requires highest level of engineering expertise from various engineering disciplines, such as chemical, mechanical, electrical engineering.

Industrial constructions are very specialized job and consists of large complex mechanical, electrical, instrumentation work.

Construction is dominated by skilled labour and quite labour intensive compared to heavy engineering but large equipments are also used.

Factory

Volkswagen factory in Wolfsburg, Germany

A factory (previously manufactory) or manufacturing plant is an industrial site, usually consisting of buildings and machinery, or more commonly a complex having several buildings, where workers manufacture goods or operate machines processing one product into another.

Factories arose with the introduction of machinery during the Industrial Revolution when the capital and space requirements became too great for cottage industry or workshops. Early factories that contained small amounts of machinery, such as one or two spinning mules, and fewer than a dozen workers have been called "glorified workshops".

Most modern factories have large warehouses or warehouse-like facilities that contain heavy equipment used for assembly line production. Large factories tend to be located with access to multiple modes of transportation, with some having rail, highway and water loading and unloading facilities.

Factories may either make discrete products or some type of material continuously produced such as chemicals, pulp and paper, or refined oil products. Factories manufacturing chemicals are often called *plants* and may have most of their equipment – tanks, pressure vessels, chemical reactors, pumps and piping – outdoors and operated from control rooms. Oil refineries have most of their equipment outdoors.

Discrete products may be final consumer goods, or parts and sub-assemblies which are made into final products elsewhere. Factories may be supplied parts from elsewhere or make them from raw materials. Continuous production industries typically use heat or electricity to transform streams of raw materials into finished products.

The term *mill* originally referred to the milling of grain, which usually used natural resources such as water or wind power until those were displaced by steam power in the 19th century. Because many processes like spinning and weaving, iron rolling, and paper manufacturing were originally powered by water, the term survives as in *steel mill*, *paper mill*, etc.

Reconstructed historical factory in Žilina (Slovakia) for production of safety matches. Originally built in 1915 for the business firm *Wittenberg and son.*

History

Max Weber considered production during ancient times as never warranting classification as factories, with methods of production and the contemporary economic situation incomparable to modern or even pre-modern developments of industry. In ancient times, the earliest production limited to the household, developed into a separate endeavour independent to the place of inhabitation with production at that time only beginning to be characteristic of industry, termed as "unfree shop industry", a situation caused especially under the reign of the Egyptian pharaoh, with slave employment and no differentiation of skills within the slave group comparable to modern definitions as division of labour.

Entrance to the Venetian Arsenal by Canaletto, 1732.

Interior of the Lyme Regis watermill, UK (14th century).

According to translations of Demosthenes and Herodotus, Naucratis was a, or the only, factory in the entirety of ancient Egypt. A source of 1983 (Hopkins), states the largest factory production in ancient times was of 120 slaves within 4th century BC Athens. An article within the New York Times article dated 13 October 2011 states:

"In African Cave, Signs of an Ancient Paint Factory" - (John Noble Wilford)

... discovered at Blombos Cave, a cave on the south coast of South Africa where 100,000-year-old tools and ingredients were found with which early modern humans mixed an ochre-based paint.

Although The *Cambridge Online Dictionary* definition of factory states:

a building or set of buildings where large amounts of goods are made using machines

elsewhere:

... the utilization of machines presupposes social cooperation and the division of labour

—*von Mises*

The first machine is stated by one source to have been traps used to assist with the capturing of animals, corresponding to the machine as a mechanism operating independently or with very little force by interaction from a human, with a capacity for use repeatedly with operation exactly the same on every occasion of functioning. The wheel was invented *circa* 3000 BC, the spoked wheel c.2000 BC. The Iron Age began approximately 1200-1000 BC. However, other sources define machinery as a means of production.

Archaeology provides a date for the earliest city as 5000 BC as Tell Brak (Ur *et al.* 2006), therefore a date for cooperation and factors of demand, by an increased community size and population to make something like factory level production a conceivable necessity.

According to one text the water-mill was first made in 555 A.D. by Belisarius, although according to another they were known to Pliny the Elder and Vitruvius in the first century B.C. By the time of the 4th century A.D. mills with a capacity to grind 3 tonnes of cereal an hour, a rate sufficient to meet the needs of 80,000 persons, were in use by the Roman Empire.

The Venice Arsenal provides one of the first examples of a factory in the modern sense of the word. Founded in 1104 in Venice, Republic of Venice, several hundred years before the Industrial Revolution, it mass-produced ships on assembly lines using manufactured parts. The Venice Arsenal apparently produced nearly one ship every day and, at its height, employed 16,000 people.

Industrial Revolution

Cromford mill as it is today.

One of the earliest factories was John Lombe's water-powered silk mill at Derby, operational by 1721. By 1746, an integrated brass mill was working at Warmley near Bristol. Raw material went in at one end, was smelted into brass and was turned into pans, pins, wire, and other goods. Housing was provided for workers on site. Josiah Wedgwood in Staffordshire and Matthew Boulton at his Soho Manufactory were other prominent early industrialists, who employed the factory system.

The factory system began widespread use somewhat later when cotton spinning was mechanized.

Richard Arkwright is the person credited with inventing the prototype of the modern factory. After he patented his water frame in 1769, he established Cromford Mill, in Derbyshire, England, significantly expanding the village of Cromford to accommodate the migrant workers new to the area. The factory system was a new way of organizing labour made necessary by the development of machines which were too large to house in a worker's cottage. Working hours were as long as they had been for the farmer, that is, from dawn to dusk, six days per week. Overall, this practice essentially reduced skilled and unskilled workers to replaceable commodities. Arkwright's factory was the first successful cotton spinning factory in the world; it showed unequivocally the way ahead for industry and was widely copied.

Between 1820 and 1850 mechanized factories supplanted traditional artisan shops as the predominant form of manufacturing institution, because the larger-scale factories enjoyed a significant technological advantage over the small artisan shops. The earliest factories (using the factory system) developed in the cotton and wool textiles industry. Later generations of factories included mechanized shoe production and manufacturing of machinery, including machine tools. Factories that supplied the railroad industry included rolling mills, foundries and locomotive works. Agricultural-equipment factories produced cast-steel plows and reapers. Bicycles were mass-produced beginning in the 1880s.

The Nasmyth, Gaskell and Company's Bridgewater Foundry, which began operation in 1836, was one of the earliest factories to use modern materials handling such as cranes and rail tracks through the buildings for handling heavy items.

Large scale electrification of factories began around 1900 after the development of the AC motor which was able to run at constant speed depending on the number of poles and the current electrical frequency. At first larger motors were added to line shafts, but as soon as small horsepower motors became widely available, factories switched to unit drive. Eliminating line shafts freed factories of layout constraints and allowed factory layout to be more efficient. Electrification enabled sequential automation using relay logic.

Assembly Line

Factory Automation with industrial robots for palletizing food products like bread and toast at a bakery in Germany.

Henry Ford further revolutionized the factory concept in the early 20th century, with the innovation of the mass production. Highly specialized laborers situated alongside a series of rolling ramps would build up a product such as (in Ford's case) an automobile. This concept dramatically decreased production costs for virtually all manufactured goods and brought about the age of consumerism.

In the mid- to late 20th century, industrialized countries introduced next-generation factories with two improvements:

1. Advanced statistical methods of quality control, pioneered by the American mathematician William Edwards Deming, whom his home country initially ignored. Quality control turned Japanese factories into world leaders in cost-effectiveness and production quality.
2. Industrial robots on the factory floor, introduced in the late 1970s. These computer-controlled welding arms and grippers could perform simple tasks such as attaching a car door quickly and flawlessly 24 hours a day. This too cut costs and improved speed.

Some speculation as to the future of the factory includes scenarios with rapid prototyping, nanotechnology, and orbital zero-gravity facilities.

Historically Significant Factories

Highland Park Ford plant, c. 1922

- Venetian Arsenal
- Cromford Mill
- Lombe's Mill
- Soho Manufactory
- Portsmouth Block Mills
- Slater Mill Historic Site
- Lowell Mills
- Springfield Armory
- Harpers Ferry Armory
- Nasmyth, Gaskell and Company also called the Bridgewater Foundry

- Baldwin Locomotive Works
- Highland Park Ford Plant
- Ford River Rouge Complex
- Hawthorne Works

Siting the Factory

A factory worker in 1940s Fort Worth, Texas, United States.

Before the advent of mass transportation, factories' needs for ever-greater concentrations of laborers meant that they typically grew up in an urban setting or fostered their own urbanization. Industrial slums developed, and reinforced their own development through the interactions between factories, as when one factory's output or waste-product became the raw materials of another factory (preferably nearby). Canals and railways grew as factories spread, each clustering around sources of cheap energy, available materials and/or mass markets. The exception proved the rule: even greenfield factory sites such as Bournville, founded in a rural setting, developed its own housing and profited from convenient communications systems.

Regulation curbed some of the worst excesses of industrialization's factory-based society, a series of Factory Acts leading the way in Britain. Trams, automobiles and town planning encouraged the separate development of industrial suburbs and residential suburbs, with laborers commuting between them.

Though factories dominated the Industrial Era, the growth in the service sector eventually began to dethrone them: the focus of labor in general shifted to central-city office towers or to semi-rural campus-style establishments, and many factories stood deserted in local rust belts.

The next blow to the traditional factories came from globalization. Manufacturing processes (or their logical successors, assembly plants) in the late 20th century re-focussed in many instances on Special Economic Zones in developing countries or on maquiladoras just across the national boundaries of industrialized states. Further re-location to the least industrialized nations appears possible as the benefits of out-sourcing and the lessons of flexible location apply in the future.

Governing the Factory

Much of management theory developed in response to the need to control factory processes. Assumptions on the hierarchies of unskilled, semi-skilled and skilled laborers and their supervisors and managers still linger on; however an example of a more contemporary approach to handle design applicable to manufacturing facilities can be found in Socio-Technical Systems (STS).

Shadow Factories

A shadow factory is a term given to dispersed manufacturing sites in times of war to reduce the risk of disruption due to enemy air-raids and often with the dual purpose of increasing manufacturing capacity. Before World War II Britain had built many shadow factories.

British Shadow Factories

Production of the Supermarine Spitfire at its parent company's base at Woolston, Southampton was vulnerable to enemy attack as a high-profile target and was well within range of *Luftwaffe* bombers. Indeed, on 26 September 1940 this facility was completely destroyed by an enemy bombing raid. Supermarine had already established a plant at Castle Bromwich; this action prompted them to further disperse Spitfire production around the country with many premises being requisitioned by the British Government.

Connected to the Spitfire was production of its equally important Rolls-Royce Merlin engine, Rolls-Royce's main aero engine facility was located at Derby, the need for increased output was met by building new factories in Crewe and Glasgow and using a purpose-built factory of Ford of Britain in Trafford Park Manchester.

Basic Terminology in Project Management

Planning:

- Predetermined course of action to obtain the desired project activities. It also involves the formulation of alternatives and chooses the best suitable. During planning phase, project is required to be split up into well definable work in terms of time and resources.

Scheduling:

- It is the arranging of the work plan in the sequence of time.

Scoping:

- Scoping is the defining the object of the project. Thus, the project must be stated in the definite end measurable terms, which include costs, schedule or quality of performance. Thus we obtain realistic and specific objective of the project during scoping.

Organizing:

- Integrating the resources in the project is termed as organizing. Money, equipment, time, technology, manpower, materials are resources. Putting them together to obtain desired result in organizing.

Controlling:

- Controlling is the monitoring, influencing and directing the project.
- The progress of project is assessed through suitable measure; deviations are discovered and corrective measures are taken to rectify the deviation.

Following steps are involved in controlling a project.

- Assess the progress of the project, measure quality and flow of money.
- Compare with original schedule, specification, design. Also obtain the quality of conformance.
- Take necessary action to correct if there is any deviation from the earlier plans.

Directing:

- It is function of project leader to give instructions to subordinates, supervise their work and corresponds to the report when they give as feedback. Basically directing is linked with organization setup of project management team.

Staffing:

- This is the process of dividing the project work and placing suitable person at the appropriate positions. Recruitment, training and their assessment are part of staffing.

Coordinating:

- This is the process of interaction between different departments or functionalities in the project management group which generally follows pyramidal structure.

Estimation:

- Estimation evaluates the resources in terms of cost. Planning evaluates the resources in terms of time. Adding both together as is necessary to obtain the cash flow.

Life Cycle of a Construction Engineering Project

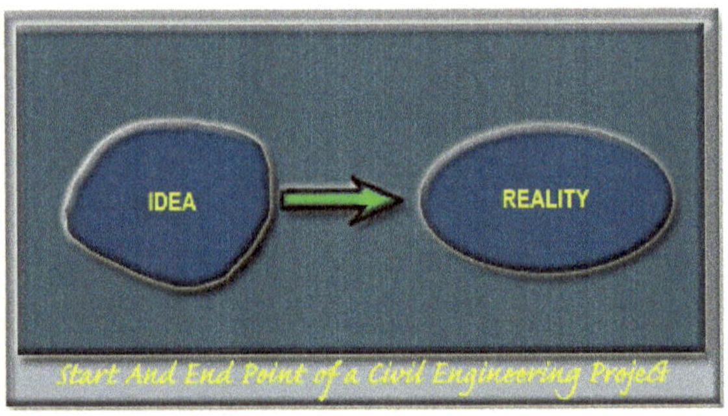

Civil Engineering project starts with idea, requirement to fulfill demands, will to provide basic amenities to the citizen by the state. Project ends with the infrastructure of facilities, recreational amenities, industrial plants in operation as reality.

The phases through which a civil engineering project passes are:

1. Conceptual and Feasibility Studies.
2. Engineering Design.
3. Contracting and Procurement
4. Construction.
5. Commissioning of a project Start-up and Implementation
6. Utilization and Maintenance.

Every civil engineering project is unique in its character and varies with respect to time and degree of emphasis with respect to the aforementioned phases during its life cycle but these projects certainly passes through these phases. A bar-chart is shown for a typical project is show in the next slide.

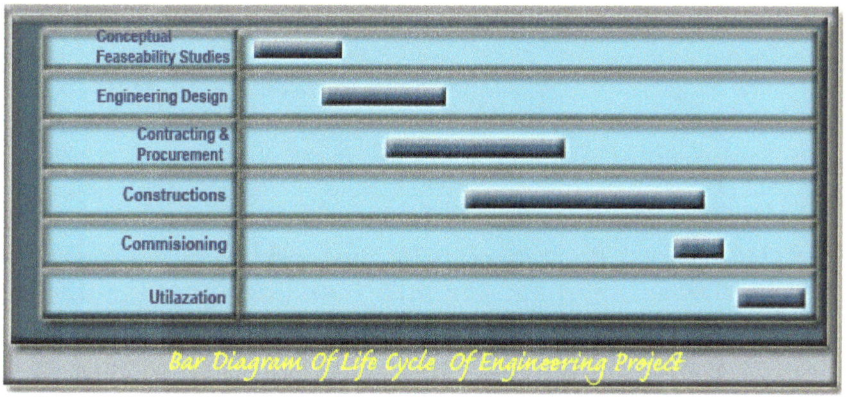

Depending upton the contractual relations and way of project management, design and construction can overlap to the various degrees

Conceptual and Feasibility Studies- In this stage, ideas generated by indivviduals / a group of people are studied. Following important points are carried out to establish the viability of the project.

- Conceptual development of the model and its studies
- Technical and economic feasibility studies
- Enviromental impact assesment
- Social Impact assesment
- Land and geological survey-location of the project
- Enumeration of major problems in translating the project to reality

Example: Suppose a state highway is crossing the railway line. The idea is to construct a fly over . This is situated in a small district level town. Here the project is the construction of the flyover. In this phase, the following points are examined-

a. Based upton the level of traffic, the basic model of the fly over is prepared, with respect to width, length of the approach road, etc.

b. The cost of construction is roughly estimated and source of money discussed.

c. Is it necessary to have a fly over now ? This point is studied with regard to the level of traffic on the road as well as on the railway line. If the traffic load is low, the project may be discarded based on cost-benifit analysis.

d. How to relocate any power line comming in the construction area.

e. Preliminary soil investigation

These points basically come under the conceptual model & technical and educational studies. Environment project assesment is not very important for this project. But a thorough study of the environment project impact has also to be carried out for the construction of Dam, Nuclear power plant, etc. In out country many of bitg dams such as Narmada and Tehri Dam has been delayed due to the object of enviromental impact. In the USA, many of the Nuclear power plants were discarded due to the enviromental assesmsent or either got converted to fossil fuel plants.

Outcome of this stage: A report dealing with investigation and studies concerning the problems to be encountered and its solution.

- Engineering Design - In this stage, project takes the definite shape. All the technical issues related to the project are carried out. Engineers come out with the best alternatives available.

  Design phase of project is carried out in two phases -

  - Preliminary design
  - Detailed design
  - Preliminary design This stage involves primarily the studies of various design alternative, their economic comparitive studies and architectural aspect. Also, in this stage, detailed field investigation such as soil testing, geological, hydrological data collection, market survey, etc are carried out. Depending on the nature of the project, various data is collected. It is difficult to say the end of feasibility study and the start of preliminary design. These steps are also a part of feasibility study.

    Example:

    In the example of a flyover, this steps involve selection of the bridge deck type, i.e. wheather a steel or concrete, boxgrider bridge or prestress. Also the exact dimension of the flyover is fixed, i.e.

  number of piers, width of foot path, architectural aspect, etc. Also detailed soil investigation and site investigation are carried out to get the idea of method of construction.

- Detailed Design- In this stage, the project is broken down into its components and each of the elements are analysed and designated, such that the engineer renders the explicit drawings and specifications which is used for construction.

This phase is traditionally involved with designers.

Outcome of this design phase : The following are the outcome of this phase -

- Drawing of the structures/facilities to be constructed
- Specifications of component to be employed which contain detailed description of the amenities to be constructed

- Contracting and Procurement In this stage, the bill of quantities are prepared, final cost estinate is worked out. Most of the agreements are signed and relationships between different parties involved is defined following points are generally addresed during this time. The project has to be executed according to the drawing and specification.Also -

  - Tender notices are advertised
  - Qualification of contractor / suppliers is studied.
  - Contract is awarded to execute the work
  - For many of the projects, such as industrial projects, industrial material and equipment to be installed is purchased

  Outcome of this stage: Contract document, which is a legal document describing the terms and conditions to execute the project.

The figure shows the life cycle of an engineering project

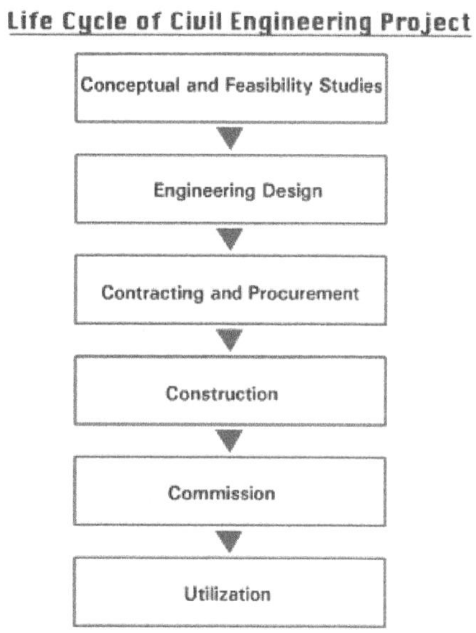

- Construction -This is the most important phase of a civil engineering project, where the major portion of the money is invested. A lot of issues of project management are invloved during this phase. The project is realized in the real world. The main work involved during construction is to answer the following questions -

  - How to do the particular construction (Construction Methods)
  - How to organize labour and equpmetns to accomplish the construction (Labour and equipments Management)
  - How to manage materials for constructions (Material Management)
  - How to breakdown the overall work into small identifiable work called activity and put them in sequence (Construction Planning)

  - How to calculate the time and resources required to complet the activity and obtain the criricality of the activity (Construction Scheduling)
  - How to distribute the labour and equipments, in order to accomplish the work in the given time (Labour, equipment, allocation, resource allocation)

1. How to prepare the bill for the money invested in the construction to get the payment from the client (Construction Finance)
2. How to conform the construction is according to drawing the specifications (Quality of conformance)
3. How to implement the safety measure taken during construction (Construction safety)
4. How to organize the different personals or the company involved in the construction and define relations among them (Constructional Organization management, construction sub contracting, construction Supervision and method of reporting)
5. How to check whether the project is accroding to construction schedule and quality (Construction supervision)
6. What to do if things are not moving according to plan (Disputes anr arbitration, Labour laws, Penalty)
7. How to avoid redundancy in the system (Value engineering)

   Outcome of this stage : Facility or structures

- Commissioning of a project - Aim of this phase is to check that the project has been carried out according to the contract document. All the facilities developed should function properly as envisaged in the design. Also during this stage changes in the design during the construction is properly documented for the purpose of technical performance and financial implications. Following activities are carried out during this stage -
  - Check if each component is functioning according to design and specification
  - Check if the entire system is functioning accordingly as desired
  - To remove the defect if any during construction work
  - Dream has come to reality

    This is one of the main stage, in the bigger project such as power plant, industrial project, dam, etc, The commissioning in large project is carried out in several stages and then the complete system is tested. Testing of components are generally carried out during construction phase. Operational and maintanance instruction manuals are also prepared during this phase.

- Utilization and maintanance -
  - Basic infrastructure facilities created should be utilized to its fullest capacity
  - Regular repair of the part gettign damaged has to be accomplished

This stage has become a very important stage in dealing with project when we talk about the life cycle cost of the project.

References

- "Traffic-related Air Pollution near Busy Roads". American Journal of Respiratory and Critical Care Medicine Vol 170. pp. 520-526. 2004
- Lay, Maxwell G (1992). Ways of the World: A History of the World's Roads and of the Vehicles that Used Them. Rutgers University Press. ISBN 0-8135-2691-4
- "Architects / Building Engineers in Japan" (PDF). The Japan Architectural Education and Information Center. p. 5. Retrieved 2014-08-16
- O'Flaherty, Coleman A. (2002). Highways: The Location, Design, Construction & Maintenance of Road Pavements. Elsevier. ISBN 0-7506-5090-7
- C. Michael Hogan, "Analysis of highway noise", Journal of Water, Air, & Soil Pollution, Volume 2, Number 3, Biomedical and Life Sciences and Earth and Environmental Science Issue, Pages 387-392, September, 1973, Springer Verlag, Netherlands ISSN 0049-6979
- Victor E. Saouma. "Lecture notes in Structural Engineering" (PDF). University of Colorado. Retrieved 2007-11-02
- Ron Bridle; John Porter (2002). The Motorway Achievement: Frontiers of Knowledge and Practice. Thomas Telford. p. 252. ISBN 0-7277-3197-1
- "Some Useful Numbers on the Engineering Properties of Materials (Geologic and Otherwise)" (PDF). Stanford University. Retrieved 2013-12-05
- O'Flaherty, Coleman A. (2002). Highways: The Location, Design, Construction & Maintenance of Road Pavements. Elsevier. p. 252. ISBN 0-7506-5090-7
- "ETABS receives "Top Seismic Product of the 20th Century" Award" (PDF). Press Release. Structure Magazine. 2006. Retrieved April 20, 2012
- G. Allen Burton, Jr.; Robert Pitt (2001). Stormwater Effects Handbook: A Toolbox for Watershed Managers, Scientists, and Engineers. New York: CRC/Lewis Publishers. ISBN 0-87371-924-7
- "Road Infrastructure Strategic Framework for South Africa". A Discussion Document. National Department of Transport (South Africa). Retrieved 2007-03-24
- Kincaid, Peter (1986). The Rule of the Road: An International Guide to History and Practice. Greenwood Press. ISBN 0-313-25249-1
- "The Somerset Levels (the oldest timber trackway discovered in Northern Europe)". Current Archaeology 172. Current Archaeology. February 2001. Archived from the original on 2007-03-11. Retrieved 2007-03-25
- John Noble Wildord (1994-05-08). "World's Oldest Paved Road Found in Egypt". New York Times. Retrieved 2012-02-11
- Bozorgnia, Yousef; Bertero, Vitelmo V. (2004). Earthquake Engineering: From Engineering Seismology to Performance-Based Engineering. CRC Press. ISBN 978-0-8493-1439-1
- "Horses and Canals 1760 - 1960 The people & the horses". Horse Drawn Boats. Canal Junction Ltd. Retrieved 2007-04-09
- Hart-Davis, Adam (2001-06-01). "Roads and surveying". Discovering Roman Technology. BBC.CO.UK. Retrieved 2007-04-22
- Lindeburg, Michael R.; Baradar, Majid (2001). Seismic Design of Building Structures. Professional Publications. ISBN 1-888577-52-5

2

Contract Management: An Overview

A civil engineering project is shared among the owner, the designer and the contractor. Good communication among these members and various others involved is necessary for a successful project. Delivery methods or contractual relation and types of contract deal with the various forms of communication. The major categories of types of contracts and contract management are dealt with great details in the chapter.

Contractual Relations

During the life cycle of projects, various parties come together and work to fulfill objective of the project. We need to know how these parties are communicating to each other and what ways are they communicating among themselves. The first point is called delivery methods or contractual relations and the second issues are termed as type of contracts. The flow of information between various parties involved takes place in many ways. i.e . there are different approaches used to organize the project team to manage the entire design and construction process. Thus we shall discuss various delivery methods. The relations existing between various parties also vary in terms of monetary payment and goods as well as service delivery. These points we shall discuss for type of contracts.

The owner needs to know how to communicate with contractor or designer. Designer or contractor can direct report to him or owner can hire a person or company which can help him in fixing, communicating designer, contractor. There are other many possibility such that he can have in-house specialists who can help him in designing or supervising the work. Similarly, owner can pay the contractor or designer based on unit work of execution or percentage of the total cost based on the contractual agreements. In this chapter, we shall discuss the various arrangements between project team and contractual agreements existing between them. We shall also discuss the process of awarding contracts, documents involved.

Although the exact nature of arrangement in terms of reporting sub reporting getting feed-back, payments of execution is unique to the particular project, these discussions will help us in understanding particular project team arrangement. Also, we can make our own arrangement depending upon the local nature of the project.

From the degree of freedom point of view we can classify the project as:

a. Public project.

b. Private project.

Public Project

Public project consists of primary infrastructure project. Public projects work on fixed sets of guideline fixed by particular department. For example in India, Central Public Works Departments (CPWD) works with their own rules. These are called CPWD manuals. These manuals are well documented and describe what to do when and how. Deviation from these documents needs special permission from apex body depending on the nature of change or sometimes there is almost no scope of change.

Similarly each state government maintains its own contract rules to execute the work. Like Assam public works departments has its own way of doing the things. Similarly Indian railways execute the works based on guidelines fixed by the department. Similar situation are existing in other countries. As for example, US transport department does the work according to its own document. But, there are many points common between their contractual arrangement as well as contractual agreements.

Contracts are selected based on competition bidding. In our country, public project mainly works on minimum cost bidding system. In recent year, much effort has been paid to improve the low cost bidding system by introducing technical bidding and price bidding separately.

Private project: There are more flexibilities in the private project. Depending on the nature of the project, the owner or project manager can arrange the project team. Whom to select, how to select and what kind of information exchange systems will exists between various project team members is dependent on the owner. Due to these natures, various forms of delivery methods have developed in practice. Owner select the designer or contractor based on the quality and reputation of the contractor plus the negotiated money or competition bidding.

Contractual Relationship or Delivery Methods

We are going to talk about how different member during of project team gets organized during the life cycle of the project or in the process of design and construction in particular, or life cycle of project. The different approaches are called Delivery methods. In general, the relations existing in a particular form follows specific pattern. But we shall discuss major arrangement followed in the construction industry. The following delivery methods cover wide arrange of contractual arrangements

a. Traditional Methods

b. Design- construct

c. Project management method

Traditional Method: In this method, owner acts as overall coordinator of the project. He engages a designer or design form to design the facilities and also prepare complete set of contract documents. With complete set of contract documents, owner employs contractor either through negotiation or competitive bid. The contractor hired is totally responsible for the delivery of the project. The designer may employ other designer to do the specific work. Such as many designer takes help of other team to do the electrical,. mechanical design in the building project and prepare the structural design in-house.

The main designer works as a coordinator of the design-process. Similarly, the contractor may subcontract out some job to subcontractor. These subcontractors will be directly reporting to the main contractor. There is no formal agreement between contractor and designer. Owner himself oversees the quality control. Sometimes designer may get involved in overseeing construction work depending upon the owner need.

Thus we can see, for large project, owner needs to maintain an engineering department to hire designer and contractor and oversees the work. There are many variations of these arrangements. Owner may have their own design group and hires only designer for specific jobs or owner can subcontract the jobs himself and oversees the project, such as award the contract of piling for one company and superstructure to other company. A typical government organization in India such as Central Public Works Department (CPWD), irrigation department works with this pattern. IIT construction work is generally carried out by this method. A typical relation is shown in the figure .

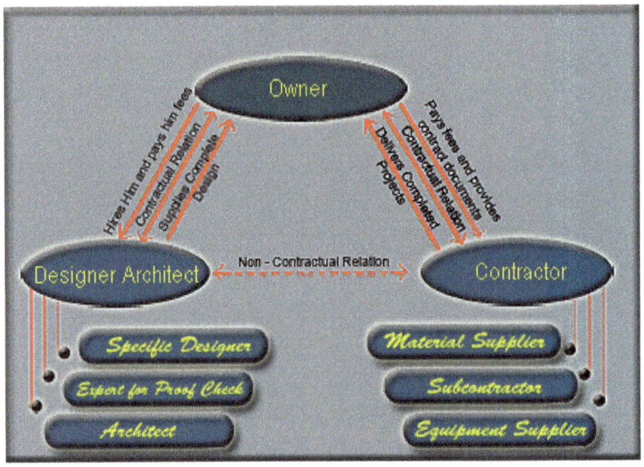

The ways fees are paid to contractor or designer depend up the type of contract. We shall discuss this point later on. It may be unit price, percent of cost or lump-sum etc. depending on the nature and agreement in the project. We shall now discuss the advantages and disadvantages of this contractual agreement.

Advantage

a. As the owner receives complete set of drawing and specification of the work, he gets the fixed price of the project. He can award the contract in any type of contract; he will have good idea of the cost of project.

b. This system is followed in public sector project. Owner, designer, contractor works in the fixed frame work. They might be working many projects. Generally well-documentations are available for different process. CPWD maintains well documented procedure.

c. Competitive biding is easy to implement as complete set of contract document is available before construction starts.

d. Owner need not to worry about day to day activity of the project. He gets involved only at the specific points.

Disadvantage

a. Contractors and subcontractors have no input in the design. Material or methods specified may not be available or contractors have knowledge of efficient methods and materials.

b. It is difficult to have overlap between design and construction phase. So fast tract construction may not be possible.

c. Sometimes there may exist different interpretations of contract document by owner and contractor.

d. The change in the methods or material of construction may give rise to conflict. Unforeseen changes are difficult to implement.

Design-Construct

In this method of delivery, design and construction is carried out by the same company. The company may be a single unit or joint venture of two companies or more than two companies. A single company can take a job for design and construction, but the company sub-contract or contract the construction, but engaged himself in the management of construction. There are different variations available for this mode of delivery. But common point is that the same company takes the job from conceptual design phase to construction and implementation. The variations of this mode are turn-key project.

Design-manage

This method of delivery has been practiced in the specialized job in industrial sector. Petrochemical industry, manufacturing plants, nuclear plants, offshore oil drilling platform. In IIT Guwahati , air-conditioning was done by "Blue Star" as design-construct methods. This method has also been used in building project recently. A typical relation is shown in figure.

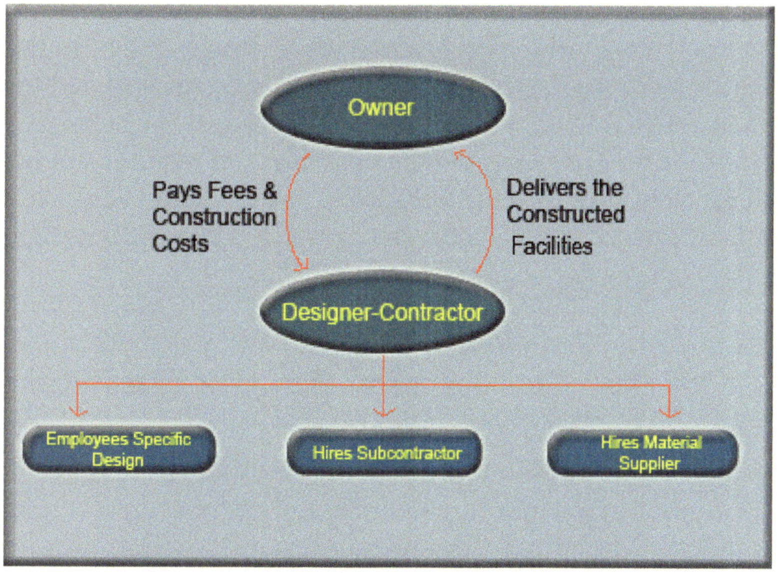

Mode of payment may vary depending on the type of construction. It may be unit price, cost plus fees, lump-sum contract. We shall discuss the advantages and disadvantages of this method of delivery in the following

Advantage

1. In this method of delivery, designers get benefit of the input from the contractor. Thus quality of design improves.

2. Due to good coordination between design and construction phase, project duration can be decreased.
3. Scope of constructability analysis and value engineering is facilitated.
4. Unforeseen changes can be easily incorporated and scope of dispute doesn't arise.

Disadvantage

1. Price of the constructed facility is not known before the construction starts.
2. Quality gets sacrificed if project is awarded as fixed price and design construct cost is in adverse situation.
3. In the case of unit price contract, inefficient form may increase the total cost of construction.
4. Owner is not aware of the construction and design. This lack of knowledge may be handicapped during maintenance phase.
5. In the traditional form, owner monitors the quality of conformance with the design document. But in design-construct method of delivery, there is no standard for the owner to check and monitor.
6. This arrangement doesn't guaranteed the best possible price.

Project Management Method

There are many variations of this mode of delivery. Some of these variations are program management; construction management, professional construction management. In this method of delivery, owner hires a construction management firm at start of the project and involves them in the project from conceptual design through implementation [start up]. The degree of involvement of the construction management firm depends upon the expertise and requirement of the owner. Owner may involve the management firm to select designer, coordinate the work between designer and contractor, monitor the constructions. A typical arrangement is shown in the figure.

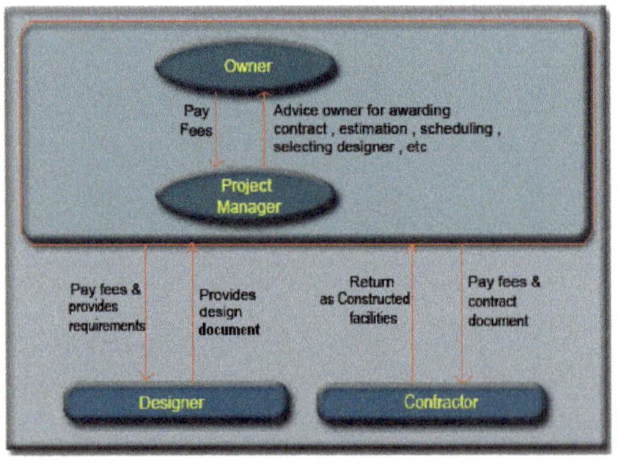

Parties Involved in Contract

We shall call the group, which is involved with construction as contractor. In the civil engineering project, in a narrow sense, ideas of owners is translated into engineering. Creation by designer/ architect and is given real force by contractor. We have a three party system as -

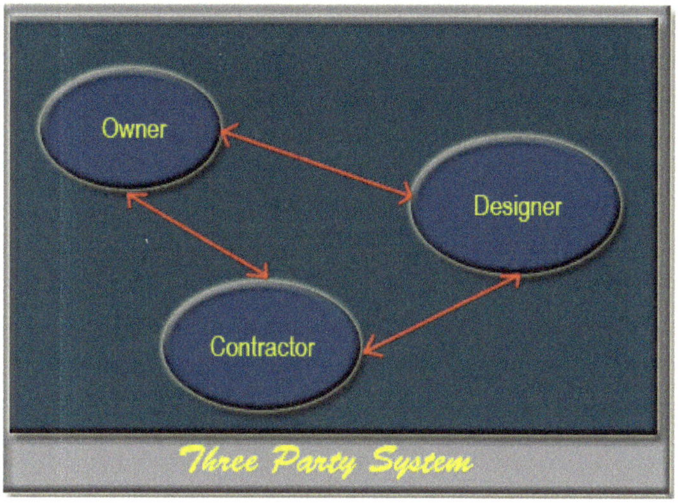

This isthe simple model for the party involved in the civil engineering construction. These three bodies are having contradictory objectives, but they work in non-adversery conditions and should help each other.

In the civil engineering project, major activities are -

a. Design

b. Construction

c. All the activity and process are to help them. Thus, we can broadly say that design is -

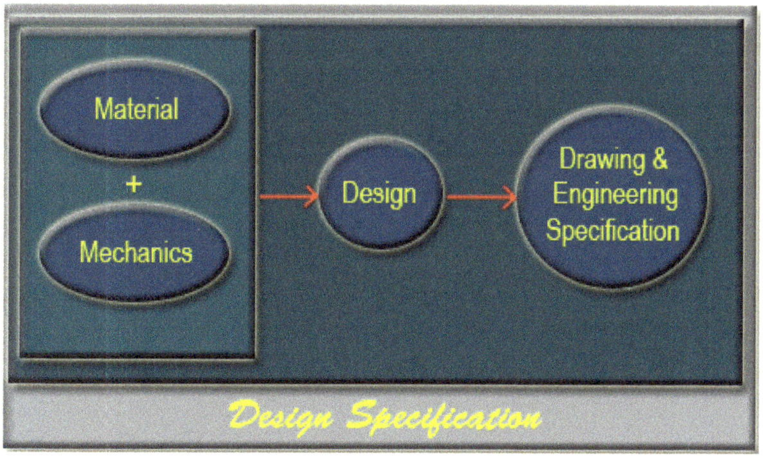

Thus, the persons involved in design are involved with the mechanics and properties of materials. We shall call the group of people involved in designs as designer. A designer may be a person or a group of persons or a company.

Other engineering creation of designer has to translate into the real facilities. This is process of construction, which is -

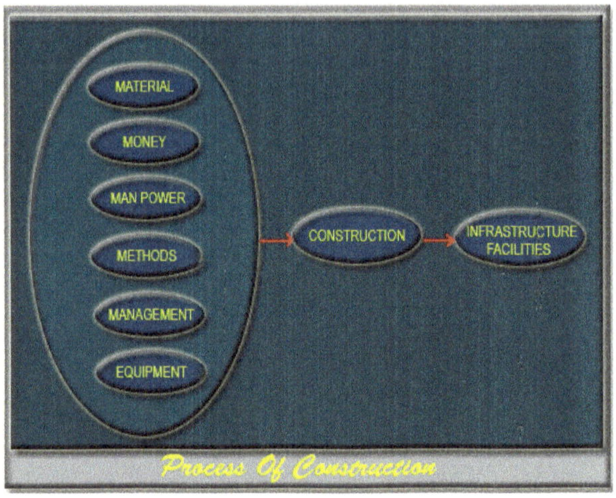

Contract

Contract is an agreement between two or than two parties in which one is agreed to provide goods and services to other for which he will get the return in the some form. The contract is legal document and binding on both the parties. The form in which return for providing goods and services is delivered, is called type of contracts. There are different types of contracts which can be employed in any of the delivery methods. Owner can pay the money to the contractor, in lump-sum, based on measured work with unit price, based on percentage plus quantity involved. In the following, we shall briefly discuss the different type of contract.

a. Lump-sum contract: This is a single fixed price contract. In this contract, contractor agrees to perform specified job for fixed sum. The owner provides the contractor exact specification of the work. In this contract both the parties try to fix the conditions of the work as precisely as possible. Following are the advantages of the fixed price contract.

   a. Owner is aware of the cost of the project before the project construction starts.
   b. It avoids a lot of details and accounting by both owner and contractor.
   c. Contractor gets free hand to execute the work.
   d. If this contract is used with design-construct method of delivery, contractor gets opportunity to use value engineering.

Disadvantage:

   a. It is very difficult to accommodate any charge in design and specification.
   b. This contract is as good as the accuracy of the contract document. If errors exist in the contract document, the contract need to be renegotiated and hence more risk is involved from the owner ride.
   c. In the case of unforeseen hazard during the construction, contractor may be put in adverse situation.

This type of construct is suited for small job, precisely specified job, low risk with construction job. This is generally suited for the job where it is easy to make the measurement. Lump-sum contract should be avoided for underground work.

Lump-sum contract with design construct method of delivery is often called turn key contract.

Payment by the owner can be carried out in lump-sum contract as total amount at time or percentage of total cost after finishing certain amount of work. For example suppose a water tank was awarded as lump-sum contract. It can be said that 15% of amount will be paid after construction of foundation 50 % of total amount will be paid after construction of staging and 80% of the total amount will be after the construction of the tank. Rest will be paid only when the tank becomes operational. Lump-sum contract in often used in sub-contracting for labour contract. In our country many laborers for excavation, plastering work with this method.

Unit Price Contract:

In this type of contract, the price is paid per unit of the work carried out. There are different variations of this type of contract. Some of them are mentioned below.

Bill of quantities contract: In this type of contract owner provide the drawing, quantities of work to be done and specification. The contractor bid based on the unit cost of the items of construction. The contractor overhead, profit and other expenses can be included in the unit cost of the item of work. Sometimes contractor quotes the unit price of the work and lump-sum amount separately as profit overhead. The estimated quantities of the work to be done called Bill of the quantities is fixed. Minor variation in the quantities is admissible in this type of contract. The drawing of the work is not suppose to change. Although change and deviations from original drawing could be accepted during construction but even then unit price does not change.

This type of construction is usually followed in government sector for large infrastructure construction. This type of contract provides owner a competitive bid. Disadvantage of the methods are:

a. Owner needs to measure the quantity of work done in the field, hence requires owner presence at the site.

    - Final price of the construction is not known precisely until last price of work is completed. If there is significant difference between the estimated quantities and the reality of the situation, owner is put in adverse situation. Mistaken quantities is called unbalanced bid . Significant unbalanced bid now considered as unethical. Schedule of rate contract: Many a time, the quantity of work to be executed is not known before. Contract is signed based on the unit cost of the item of work. Generally more items are inserted in the contract than to be executed because it becomes sometimes difficult to exactly specify all the items. There is no guarantee that all the items mentioned will be used in the construction. This type of contract are widely used in underground work, flood control and road constructions. Advantages and disadvantages of this type of contract in the same as the bill qualities contract. There are other variations of schedule of rate contract where unit price plus profit is charged as the cost which we shall discuss next.

- Cost plus contract: In this contract, the payment is made based on the work carried out plus the fee which includes overhead, profit etc. Sometimes a cap is put on the type of contract by provided maximum and minimum cost limit such as guaranteed maximum cost contract. If project cost exceed this limit, contractor is responsible for that.

Sometimes incentive clause is also included if the contractor bring the project before certain specified limit.

The advantage of this type of contract is that considerable overlap is provide between design and construction. Hence the project can be executed in the fast-tract basis. This contract is suitable for the work where it is difficult to define the task to be done before the awarding the contract.

Construction Bidding

Construction bidding is the process of submitting a proposal (tender) to undertake, or manage the undertaking of a construction project. The process starts with a cost estimate from blueprints and material take offs.

The tender is treated as an offer to do the work for a certain amount of money (firm price), or a certain amount of profit (cost reimbursement or cost plus). The tender, which is submitted by the competing firms, is generally based on a bill of quantities, a bill of approximate quantities or other specifications which enable the tenders to attain higher levels of accuracy, the statement of work.

For instance, a bill of quantities is a list of all the materials (and other work such as amount of excavation) of a project which have sufficient detail to obtain a realistic cost, or rate per described item of work/material. The tenders should not only show the unit cost per material/work, but should also if possible, break it down to labour, plant and material costs. In this way the individual who is selecting the tender will be quite confident that the tender is feasible. Bids are not only chosen on cost alone. Sometimes contractors submit lower tenders to win the contract and win the work. Either the costs that the contractor incurs are greater than the price he is charging the client (as a consequence of a lower tender determining the contract sum), and thus is likely to go insolvent, or he will claim for "loss and/or expense" due to discrepancies in the contract documents (this can be done deliberately). The lowest tender is not always a feasible tender. The lowest tender is the most likely to increase the contract sum the most throughout the course of the project.

Bid Solicitation

Bid solicitation is the process of making published construction data readily available to interested parties, including construction managers, contractors, and the public. There are several services, including government entities and private plan rooms, that allow project owners to release project details to solicit and obtain contractor bids. These services act as a gateway for project owners to release project information to a large group of contractors, general contractors or subcontractors in an attempt to solicit bids. Many of these services are subscription based or charge a flat rate for project data.

Contractual Formation

Depending upon the language in the bid proposal, a subcontracting construction company could make its bid final, and, if accepted, a legally enforceable contract is created. In these circumstances, upon determination by the general contractor that a bid is the lowest offer, it can accept the bid and, upon acceptance, a subcontractor cannot renege or revoke its offer. The language of the bid or offer can impact the court's determination of whether the subcontractor intended for further negotiations to take place, or whether the bid was intended to be an option or unilateral agreement to enter into a contract upon acceptance of the offer.

Types of Project Delivery

The most common methods of construction project delivery include design-bid-build (DBB), the design-build (DB), the construction-manager-as-constructor approach and a negotiated approach. Each of these methods have advantages and disadvantages and they all can be used to successfully plan, design and undertake a given construction project.

Traditional Procurement

The traditional procurement method is the most common construction delivery method. This process begins with an owner selecting an architect to prepare construction documents. These are prepared using drafting standards such as the Institute of Civil Engineers ICE Conditions of Contract, or the NEC Engineering and Construction Contract. In most cases, the architect will release these construction documents publicly, or to a select group of general contractors, who will then place a bid on the project which reflects what they believe cost of construction will total. This bid is inclusive of a multitude of subcontractor bids for each specific trade. The general contractor's fee is generally built into the bid cost. Most government contracts are bid competitively using this method.

Digital Procurement

The digital procurement method is rapidly emerging. There are various web sites that provide electronic bidding, tender calls, scope work,designbids and related services.

Design-build

Design-build (or design/build, and abbreviated D-B or D/B accordingly) is a construction project delivery system where, in contrast to traditional "design-bid-build" (or "design-tender"), the design and construction aspects are contracted for with a single entity known as the design-builder or design-build contractor. The design-builder is usually the general contractor, but in many cases it is also the design professional (architect or engineer). This system is used to minimize the project risk for an owner and to reduce the delivery schedule by overlapping the design phase and construction phase of a project. Where the design-builder is the contractor, the design professionals are typically retained directly by the contractor.

History

The design/build delivery system often cites the original "master builder" model used to build most pre-modern projects. Under the master builder approach, a central figure of the architect held total project accountability. From inception to completion, the master builder was the key organizational figure and strictly liable to the owner for defects, delays, and losses. The design/build system is a return to some of the fundamentals of the Master Builder approach.

Overview of Process

Design-build focuses on combining the design, permit, and construction schedules in order to streamline the traditional design-bid-build environment. This does not shorten the time it takes to complete the individual tasks of creating construction documents (working drawings and specifications), acquiring building and other permits, or actually constructing the building. Instead, a design-build firm will strive to bring together design and construction professionals in a collaborative environment to complete these tasks at the same time.

Typically the hallmark of a design/build project is that one organization is responsible for both design and construction of the project. If this organization is a contractor, the process is known as "contractor-led design-build". If the organization is a design firm, the process is known as "design-led design-build". In either case, the organization employed by the owner rarely handles both aspects of design and construction in-house. In fact, the organization often subcontracts with on-site personnel (if design-led) as well as architects and engineers (if contractor-led).

Potential Problems of Design-build

Cost estimating for a design-build project is sometimes difficult because design documents are often preliminary and may change over the course of the project. As a result, design-build contracts are often written to allow for unexpected situations without penalizing either the design-builder or the owner. Several organizations (such as the Design/Build Institute of America) provide standardized form contracts for design-builders to use, but it is not unusual for the design-builder to provide its own contractual documents.

This uncertainty requires the owner to rely a great deal on the integrity, acumen, and competence of the design-builder. As the certainty of estimates decreases, the opinion of the construction professionals of the design-build firm must be trustworthy, accurate, and reasonably verifiable in order to minimize risk.

Benefits of Design-build

It is important to note that the design-build method, while not focused on saving the owner construction costs, nonetheless often saves the owner money on the overall project. The combined effects of carrying a construction loan (which typically carries a higher interest rate than permanent financing) and an earlier useful on-line date usually yields considerable overall profitability to the project and may make seemingly unfeasible projects into genuine opportunities.

The compression of time is only one important aspect of implementing this system.

Other Attributes Include

- Increased accountability by the service provider,

- Single source project delivery, and
- A value based project feedback system

Accountability

Rather than a parcelized level of responsibility of the classic design-bid-build, design-build provides an integrated solution for the owner or client. This moves projects away from the "finger-pointing" that is often commonplace in contemporary construction projects, and allows the owner to look to one entity with any questions or concerns.

Single Source

Instead of having several contractors and consultants, an owner has just one entity to deal with. Design revisions, project feedback, budgeting, permitting, construction issues, change orders, and billing can all be routed through the design-build firm. This single point of contact allows a certain degree of flexibility for the owner. Most design-builders will leverage that flexibility for the owner's benefit by continually refining the construction program to maximize the owner's value at the completion of the project.

Value-based Project Feedback

Typically, in order for a contractor to bid on a project, very specific details relating to the methods and materials must be given to avoid any ambiguity and to make an "apples to apples" comparison of bids. In a design-build context, the owner, the owner's other consultants, and the design-builder can work together to determine what methods and materials will maximize the owner's value. In instances where marginally more expensive materials, designs, or construction methods might yield a higher return on investment for the owner than those of lower cost, the owner is free to adjust the project's program without having to re-bid the entire project.

Construction Manager as Constructor

Under this delivery method, a construction manager is hired prior to the completion of the design phase to act as a project coordinator and general contractor. Unlike the DBB method, a construction manager is hired during the design phase, which allows the construction manager to work directly with the architect and circumvent any potential design issues before completion of the construction documents. After documents are completed, the construction manager accepts bids for the various divisions of work from subcontractors or general contractors.

Negotiated

This delivery method is similar to the design–bid–build method in that design and construction are performed by different firms. Unlike the design-bid-build approach, a general contractor and an architect are selected at the project's inception. These firms work together throughout the design phase. When design documents are complete, the final construction costs are negotiated by the general contractor through bids from subcontractors on various scopes of work.

Stages of Awarding Contract

The stage of awarding construction contract is dependent of the methods of delivery of the project. In the tradition way of delivery, contract documents are prepared after the completion of design. Generally designer helps the owner to prepare estimate of the project. If owner maintains an engineering department, an estimate is prepared by the engineering department. Specification of the items used for construction is very important from point of view of quality of design and quality of construction. Specification is generally prepared by the designers. During the design, designer may take owner advice in preparing specification of the items. The drawing and specification prepared by designer form the basis for the awarding contract.

A tender notice is advertised in the news paper depending upon the type of work. It may be in nation newspaper or local news paper. In a large project, tenders are invited from person/company from any country called the global tendering. Sometime this process is carried out in two stages. In the first stage the aim is to select the contractor who has good experience of executing the similar kind of job and have required technical competence. The process is called "pre qualification". The contract documents are sold to these pre qualified contractors. Many a time designer himself identifying the competent contractor based on his experience.

The contract document is bought by the contract or by paying some fees. Contractor prepares their offer and submit to the owner. The offer by contractor to the owner is called Tender and process is called submitting tender. Sometime owner ask to submit the tender in two separate documents. The first offer provides the specification of the item to be used. This is called "technical bid ". And price offer is given separate which is "price-bid".

After getting the tender, owner studies the tender documents, does comparatives studies and awards the contract to the lowest quoted contractor or who has given better technical deal. Sometimes contractor quotes the specifications which are higher or lower than the owner specifications. To get competitive bid, owner may provide this specification to other contractor and ask them to go for price bid. There is scope of using value engineering in this process also.

Thus the complete process of awarding contract can be summarized as follows:

1. Preparation of contract document
2. Advertising the tender, tender notice
3. Submitting Tender by contractor
4. Study of the tender by owner
5. Clarification regarding conditions or specifications
6. Preparing comparative list
7. Awarding contact

Contract Documents

Documents related to awarding contract are called contract documents, generally it contains following

i. drawing

ii. specification of items

iii. general conditions of contract

iv. general explanation

A tender is the offer by the contractor to the owner to construct and the execute the work according to the drawing, specification (provided by owner or changed by contractor) and general condition of controls.

Finally all the parties involved in the project (owner and contract) sign the legal agreement with respect to well defined intentions. If bidding is carried out with internationally, contract document consists of general conditions of site, its location local laws regarding labor etc., general condition for execution of the work etc.

Contractual Term

A contractual term is "Any provision forming part of a contract". Each term gives rise to a contractual obligation, breach of which can give rise to litigation. Not all terms are stated expressly and some terms carry less legal gravity as they are peripheral to the objectives of the contract.

Classification of Term

Condition or Warranty

Conditions are terms that go to the very root of a contract. Breach of a condition will entitle the innocent party to terminate the contract. A warranty is less imperative than a condition, so the contract will survive a breach. Breach of either a condition or a warranty will give rise to damages.

It is an objective matter of fact whether a term goes to the root of a contract. By way of illustration, an actress's obligation to perform the opening night of a theatrical production is a condition, whereas a singer's obligation to perform during the first three days of rehearsal is a warranty.

Statute may also declare a term or nature of term to be a condition or warranty. For example, the *Sale of Goods Act* 1979 (UK) s15A provides that terms as to title, description, quality, and sample (as described in the Act) are conditions save in certain defined circumstances.

Innominate Term

Lord Diplock, in *Hong Kong Fir Shipping Co Ltd v Kawasaki Kisen Kaisha Ltd*, created the concept of an innominate term, breach of which may or may not go to the root of the contract depending upon the nature of the breach. Breach of these terms, as with all terms, will give rise to damages. Whether or not it repudiates the contract depends upon whether legal benefit of the contract has been removed from the innocent party. Megaw LJ, in 1970, preferred the use of the classic categorizing into condition or warranty due to legal certainty. This was interpreted by the House of Lords as merely restricting its application in *Reardon Smith Line Ltd. v Hansen-Tangen*.

Enforceability

In general, parties can only sue for enforcement of valid contractual terms as opposed to representations or mere puffs.

Statements

Only certain statements create contractual obligations. Statements can be split into the following types:

- Puff (sales talk): If no reasonable person hearing this statement would take it seriously, it is a puff, and no action in contract is available if the statement proves to be wrong. It may also be referred to as "puffery". This is common in television commercials.
- Representation: A representation is a statement of fact which does not amount to a term of the contract but it is one that the maker of the statement does not guarantee the truth of. This gives rise to no contractual obligation but may amount to a tort, for example misrepresentation.
- Term: A term is similar to a representation, but the truth of the statement is guaranteed by the person who made the statement therefore giving rise to a contractual obligation. For the purposes of Breach of Contract, a term may further be categorized as a condition, warranty or innominate term.

Determination of Nature of a Statement

There are various factors that a court may take into account in determining the nature of a statement. These include:

- Timing: If the contract was concluded soon after the statement was made, this is a strong indication that the statement induced the person to enter into the contract. Lapse of a week within the negotiations of a car sale was held to amount only to a representation in Routledge v McKay
- Content of statement: It is necessary to consider what was said in the given context, which has nothing to do with the importance of a statement.
- Knowledge and expertise: In Oscar Chess Ltd v Williams, a person selling a car to a second-hand car dealer stated that it was a 1948 Morris, when in fact it was a 1939 model car. It was held that the statement did not become a term because a reasonable person in the position of the car dealer would not have thought that an inexperienced person would have guaranteed the truth of the statement.
- Reduction into Writing: Where the contract is consolidated into writing, previous spoken terms, omitted from the consolidation, will probably be relegated to representations. The old case of Birch v Paramount Estates Ltd provided that a very important spoken term may persist even if omitted from the written consolidation; this case concerned the quality of a residential house.

The parol evidence rule limits what things can be taken into account when trying to interpret a contract. This rule has practically ceased operation under UK law, but remains functional in Australian Law.

Implied Terms

A term may either be expressed or implied. An express term is stated by the parties during negotiation or written in a contractual document. Implied terms are not stated but nevertheless form a provision of the contract.

Terms Implied in Fact

The Privy Council established a five-stage test in *BP Refinery (Westernport) Pty Ltd v Shire of Hastings*. However, the English Court of Appeal sounded a note of caution with regard to the *BP* case in *Philips Electronique Grand Public SA v British Sky Broadcasting Ltd* in which the Master of the Rolls described the test as "almost misleading" in its simplicity.

1. Reasonableness and equitableness: The implied term must be reasonable and equitable. In Biotechnology Australia Pty Ltd v Pace, it was held a term that imposes a significant detriment or burden on the other party is unlikely to be equitable.

2. Business efficacy: The implied term must be necessary for the business efficacy of the contract. For instance, if the term simply causes the contract to operate better, that does not fit this criterion. This is the principle laid out in The Moorcock. The presiding judge created a quaint concept of an officious bystander; if the officious bystander were to propose a term and both the parties would be likely to reply with a testy "oh, of course", the term is implied.

3. Obviousness: The term is so obvious that it goes without saying. Furthermore, there must be one and only one thing that would be implied by the parties. For example, in Codelfa Construction Pty Ltd v State Rail Authority of NSW, a term regarding the inability of construction company to work three shifts a day could not be implied because it was unclear what form it would have taken. In English law, this principle was established in the case of Spring v NASDS, in the context of a trade union membership contract.

4. Clear expression: The term must be capable of clear expression. No specific technical knowledge should be required.

5. Consistency: The implied term may not contradict an express term.

The High Court of Australia has ruled that the test in *BP* applies only to formal contracts. In the case of an informal contract, where the parties have not attempted to stipulate the full terms, the courts should imply a term upon referring to the imputed intention of the parties, provided that the particular term is necessary for the effective operation of the contract. In implying terms in an informal contract, the High Court has suggested that a flexible approach is required. In a case where it is apparent that the parties have not attempted to spell out the full terms of their contract, the court should imply a term by reference to the imputed intentions of the parties if, but only if, it can be seen that the implication of the particular term is necessary for the reasonable or effective operation of a contract of that nature in the circumstances of the case. Obviousness also remains an important element in implying a term in an informal contract.

Terms Implied in Law

These are terms that have been implied into standardized relationships.

Common Law

- *Liverpool City Council v Irwin* established a term to be implied into all contracts between tenant and landlord that the landlord is obliged to keep the common areas in a reasonable state of repair.
- *Wong Mee Wan v Kwan Kin Travel Services Ltd* established that when a tour operator contracts to provide services, a term is implied that those services will be performed with reasonable duty and care.

Statutory

The rules by which many contracts are governed are provided in specialized statutes that deal with particular subjects. Most countries, for example, have statutes which deal directly with sale of goods, lease transactions, and trade practices. For example, each American state except Louisiana has adopted Article 2 of the Uniform Commercial Code, which regulates contracts for the sale of goods. The most important legislation implying terms under United Kingdom law are the Sale of Goods Act 1979, the Consumer Protection (Distance Selling) Regulations 2000 and the Supply of Goods and Services Act 1982 which imply terms into all contracts whereby goods are sold or services provided.

Terms Implied by Custom or Trade

One is generally bound by the custom of the industry that one is in. To imply a term due to custom or trade, one must prove the existence of the custom, which must be notorious, certain, legal and reasonable.

Course of Dealing

If two parties have regularly conducted business on certain terms, the terms may be assumed to be same for each contract made, if not expressly agreed to the contrary. The parties must have dealt on numerous occasions and been aware of the term purported to be implied. In *Hollier v Rambler Motors Ltd* four occasions over five years was held to be sufficient. In *British Crane Hire Corp Ltd v Ipswich Plant Hire Ltd* written terms were held to have been implied into an oral contract in which there was no mention of written terms.

Good Faith

It is common for lengthy negotiations to be written into a heads of agreement document that includes a clause to the effect that the rest of the agreement is to be negotiated. Although these cases may appear to fall into the category of agreement to agree, Australian courts will imply an obligation to negotiate in good faith provided that certain conditions are satisfied:

- Negotiations were well-advanced and the large proportion of terms have been worked out; and
- There exists some mechanism to resolve disputes if the negotiations broke down.

The test of whether one has acted in good faith is a subjective one; the cases suggest honesty, and possibly also reasonableness. There is no such implied term under UK common law: an attempt was made by Lord Denning in a series of case during the 1970s and 1980s but they are no longer considered 'good law'. European legislation imposes this duty, but only in certain circumstances.

The Unfair Terms in Consumer Contracts Regulations 1999 reg 8 renders ineffective any 'unfair' contractual term if made between a seller or supplier and a consumer. Regulation 5 of the statutory instrument further elaborates upon the concept of 'unfair', which is rather novel to English law. 'Unfair' is a term that was not individually negotiated (i.e. standard form) that "causes a significant imbalance in the parties' rights and obligations arising under the contract to the detriment of the consumer". This is not possible if the term is not contrary to 'good faith'; such as in *Director General of Fair Trading v First National Bank*, wherein the lack of a seemingly unfair interest term would leave the bank open to a very poor deal whereby no interest could be charged.

"Subject to" Contracts

Four Categories

If a contract specifies "subject to contract", it may fall into one of three categories as identified in *Masters v Cameron*:

1. The parties are immediately bound to the bargain, but they intend to restate the deal in a more formalized contract that will not have a different effect; or
2. The parties have completely agreed to the terms, but have made the execution of some terms in the contract conditional on the creation of a formal contract; or
3. It is merely an agreement to agree lacking the requisite *intention* to create legal relations, and the deal will only be binding unless and until the formalized contract has been drawn up.

Subsequent authorities have been willing to recognize a fourth category in addition to those stated in *Masters v Cameron*.

4. The parties intend to immediately bound by the terms agreed upon and expect to create a further contract as a replacement for the initial contract which will contain additional terms (if agreed upon).

Contingent Condition

If a contract specifies "subject to finance", it may impose certain obligations on the purchaser:

If the contract is silent on the level of effort required by the finance seeker (usually purchaser) to obtain finance, the finance seeker may come under an implied duty to cooperate. Furthermore, whether the finance seeker may validly claim non-fulfillment of a contingent condition, despite being genuinely satisfied with finance he or she obtained before the expiration of the contingent condition, was not decided in *Meehan v Jones*.

"Subject to finance" provisions may be also referred to as contingent conditions, which come under two categories: condition precedent and condition subsequent. Conditions precedent are conditions that have to be complied with before performance of a contract is required by both parties With conditions subsequent, parties do not need to perform the contract if a condition is not yet (such as official certification to practice in a particular course of business). The non-fulfillment of a contingent condition means that the parties are not required to perform their side of the respective bargain.

Contract Management

Contract management or contract administration is the management of contracts made with customers, vendors, partners, or employees. The personnel involved in contract administration required to negotiate, support and manage effective contracts are often expensive to train and retain. Contract management includes negotiating the terms and conditions in contracts and ensuring compliance with the terms and conditions, as well as documenting and agreeing on any changes or amendments that may arise during its implementation or execution. It can be summarized as the process of systematically and efficiently managing contract creation, execution, and analysis for the purpose of maximizing financial and operational performance and minimizing risk.

Common commercial contracts include employment letters, sales invoices, purchase orders, and utility contracts. Complex contracts are often necessary for construction projects, goods or services that are highly regulated, goods or services with detailed technical specifications, intellectual property (IP) agreements, outsourcing and international trade. Most larger contracts require the effective use of contract management software to aid administration among multiple parties.

A study has found that for "42% of enterprises...the top driver for improvements in the management of contracts is the pressure to better assess and mitigate risks" and additionally,"nearly 65% of enterprises report that contract lifecycle management (CLM) has improved exposure to financial and legal risk."

Contracts

A contract is a written or oral legally-binding agreement between the parties identified in the agreement to fulfill the terms and conditions outlined in the agreement. A prerequisite requirement for the enforcement of a contract, amongst other things, is the condition that the parties to the contract accept the terms of the claimed contract. Historically, this was most commonly achieved through signature or performance, but in many jurisdictions - especially with the advance of electronic commerce - the forms of acceptance have expanded to include various forms of electronic signature.

Contracts can be of many types, e.g. sales contracts (including leases), purchasing contracts, partnership agreements, trade agreements, and intellectual property agreements.

- A sales contract is a contract between a company (the seller) and a customer where the company agrees to sell products and/or services and the customer in return is obligated to pay for the product/services bought.

- A purchasing contract is a contract between a company (the buyer) and a supplier who is promising to sell products and/or services within agreed terms and conditions. The company (buyer) in return is obligated to acknowledge the goods / or service and pay for liability created.

- A partnership agreement may be a contract which formally establishes the terms of a partnership between two legal entities such that they regard each other as 'partners' in a commercial arrangement. However, such expressions may also be merely a means to reflect the desire of the contracting parties to act 'as if' both are in a partnership with common goals. Therefore, it might not be the common law arrangement of a partnership which by definition creates fiduciary duties and which also has 'joint and several' liabilities.

Areas of Contract Management

The business-standard contract management model, as employed by many organizations in the United States, typically exercises purview over the following business disciplines:

- Authoring and negotiation
- Baseline management
- Commitment management
- Communication management.
- Contract visibility and awareness
- Document management
- Growth (for Sales-side contracts)
- Contract compliance/governance

Change Management

There may be occasions where what is agreed in a contract needs to be changed later on. A number of bases may be used to support a subsequent change, so that the whole contract remains enforceable under the new arrangement.

A change may be based on:

- A mutual agreement of both parties to vary the contract, outside the framework of the existing contract. This would be an independent basis for changing the contract.

- A unilateral decision to vary the contract, contemplated and allowed for by the existing contract. This would normally have notice periods for fairness and often the right of the other, especially in consumer contracts, to cease the contractual relationship. Be careful that any one-way imposition of change is contractually justified, otherwise it may be interpreted as a repudiation of the original contract, enabling the other party to terminate the contract and seek damages.

- A bilateral decision to vary the contracting, within the variation or change control process outlined in the existing contract. These are often called change control provisions.

Phases of Contract Management

Contract Management can be divided into three phases namely

- pre- contract phase
- contract execution phase
- post award phase (often referred to as contract compliance/governance)

Contract Compliance/Governance

During the post-award phase, it is important to ensure that contract conditions and terms are met, but it is also critical to take a closer look for items such as unrecorded liabilities, under-reported revenue or overpayments. If these items are overlooked, margin may be negatively impacted. A contract compliance audit will often commence with an opportunity review to identify the highest risk areas. Having a dedicated contract compliance (and/or governance) program in place has been shown to result in a typical recovery of 2-4% and sometimes as high as 20%.

Current thinking about contract management in complex relationships is shifting from a compliance "management" to a "governance" perspective, with the focus on creating a governance structure in which the parties have a vested interest in managing what are often highly complex contractual arrangements in a more collaborative, aligned, flexible, and credible way. In 1979, Nobel laureate Oliver Williamson wrote that the governance structure is the "framework within which the integrity of a transaction is decided." He further added that "because contracts are varied and complex, governance structures vary with the nature of the transaction". Eriksson and Westerberg (2011); Li, Arditi, and Wang (2012); Chen and Manley (2014), and Cardenas, Voordijk, and Dewulf (2017) have hypothesized, developed and extensively tested conceptual models in which relevant project governance instruments and factors were identified and related to the performance of construction projects.

A collaborative governance framework has four components:

- A relationship management structure (how the parties work together to make both day to day operational decisions as well as strategic decisions)
- A joint performance and transformation management process designed to track the overall performance of the partnership
- An exit management plan as a controlling mechanism to encourage the organizations to make ethical, proactive changes for the mutual benefit of all the parties.
- Compliance to special concerns and regulations, which include the more traditional components of contract compliance

Contract Lifecycle Management

Contract Lifecycle Management (CLM) is the proactive, methodical management of a contract from initiation through award, compliance and renewal. Implementing CLM can lead to significant improvements in cost savings and efficiency. Understanding and automating CLM can also limit organizational liability and increase compliance with legal requirements.

The key steps in the Contract Lifecycle Management (CLM) process:

1) Capture: Step #1 for most people is to get all of their existing paper contracts in a central visible place. If you have no clear idea what contracts you have, then you have no idea what your potential risk exposure is. The next step is to capture the data from the contracts.

2) Track: Once you know where all your contracts are, the next step is to track any important data so you don't miss any milestone. This affects both old contracts and newly signed contracts.

3) Author: With tracking in place, the focus can shift to accelerating the new contract creation process. The "authoring" process is about setting up contract templates and clause libraries to empower business users and legal users, respectively, to create new contracts in a fast but controlled fashion.

4) Create: Contract creation has two tracks. First, empowering business users with self-service, wizard-based creation using intelligent templates pre-approved by legal. Second, making life easier for legal when drafting complex deals on your own paper, or reviewing and "patching" contracts negotiated on other people's paper.

5) Approve: Companies need an approval workflow process to ensure that risky contracts are reviewed and approved by the right person. Some risks need to be approved by legal, some by management, some by special departments like credit, etc. Ideally, safe deals are pre-approved.

6) Negotiate: The needs of the typical negotiator include: a checklist of issues for each type of deal; a way of marking up the paper to highlight how it's different from the original or the last draft; and a way of capturing and flagging those differences (if possible).

7) Sign: When the contract is final and agreed it needs to be signed by all relevant people, and the signed copies need to be stored. Someone also needs to validate that the final signed version was not modified in any way after execution copies were sent out.

8) Analyze: Enterprises with 100+ contracts need to analyze those contracts for potential risks, rights & obligations. Executives, Legal & Risk need scanned reports to manage common risks, and ad hoc reports for unexpected scenarios. "Fingertip access" to contract terms and data allows people to react quickly to new scenarios and deal with time-sensitive risks.

Contractor Management

Contractor management is the managing of outsourced work performed for an individual company. Contractor management implements a system that manages contractors' health and safety information, insurance information, training programs and specific documents that pertain to the contractor and the owner client. Most modern contracts require the effective use of contract management software to aid administration between multiple parties.

Risk and Control

Risk increases with the loss of control from outsourcing work. Keeping work in-house gives an

Owner Client complete control over the production or services provided including quality, durability, and consistency. Outsourcing the work reduces the amount of control held over these aspects. While contracts and agreements can be set in place to control the end product, the Owner Client cannot have complete assurance that their requirements are being met.

With the continuing outsourcing of production, companies struggle to standardize their contractor management processes. Requirements and regulations from the U.S. Occupational Safety and Health Administration and other governing bodies are constantly changing. Companies need to have full visibility into the quality of work their hired contractors have performed in the past and are performing now, and this often proves difficult.

There are tools which may measure the contractor's level of performance. For example, many large refineries have integrated their gate access control system to a contractor management software. This provides real-time access to the performance of the contractor workforce within the refinery.

Components

Effective contractor management relies first on a standardized prequalification form (PQF). A quality prequalification form will also allow for customized functionality, as needed. The prequalification form ensures that the necessary steps are in place for a contractor to work safely and sustainably, prior to establishing an agreement, or allowing a vendor to come on-site.

A pre-qualification form (or an explanation of the requirements) is provided before bidding or quoting to assure and include the requirements in workplans and budgets.

Specifically, the prequalification form will allow the organization to track the most importance aspect of contractor management – contractor prequalification across these essential dynamics:

- Financial stability
- Regulatory citation history
- Safety and Health statistics and programs
- Environmental protection programs
- Background checks and Security programs
- Sustainability/ Social Responsibility background and programs, including Human Rights.
- Site-specific requirements (as needed)
- Major projects performed, including references
- The length of time the contractor has been in business
- Services performed, and a risk ranking based on the contractor's trade
- Insurance coverage and limits, additional insured, and waiver of subrogation

A thorough prequalification form with each of these components is used to verify incidence rates and ensures that the contractor's insurance certification is in line with company requirements.

The prequalification form is then reviewed for OSHA logs and Experience Modification Ratings (EMR) to unearth any inconsistencies and to verify the contractor license status. Finally, references are contacted to provide actual work history and experience to further certify that the contractor is prequalified for performing work at that location.

Auditing

Once the prequalification form has been filled out, the contractor must be monitored for compliance, this is a significant part of contractor management. An audit is conducted based on the services that are performed by the contractor and the risk associated with that service.

For contractors involved in a higher risk trade (e.g. electricians, lockout-tagout and confined space workers) the audit is an essential part of reviewing the contractor's safety program. Many of these high risk workers are involved in potentially life-threatening situations and must demonstrate their ability to protect employees and their clients from harmful situations.

In order to determine that the contractor possesses an adequate understanding of both the work to be performed and the safety standards that must be followed, the following questions are asked:

1. Is the Safety Program that is currently in place adequate?
2. Has the written program Safety Program been implemented?
3. Is the inspection of critical pieces of equipment being done?
4. Does the program allow for customization in the program based on risk?
5. Is the Safety Program specific to the contractor and the services provided?
6. Are the essential programs addressed, based on the services that will be performed?
7. Is there a method to ensure the training based on written programs is being conducted?
8. Are Job Hazard Analyses (JHAs), job-site inspections and other hazard identification / mitigation techniques being used and documented?

These questions are then used as indicators to determine the contractor's level of performance. The audit ensures that the contractor has prepared for, and is enforcing, safety guidelines in their everyday practices. Essentially, the audit provides a third-party confirmation that the information supplied in the prequalification form is both accurate and up-to-date.

Database

A database can be used to record and access the vendor data within a contractor management program. The database needs to be updated regularly, to ensure that all stakeholders are kept informed of any changes, particularly if the contractor management program is being used to eliminate subpar performers. The use of an online contractor management database facilitates the sharing of contractor data in a secure format with all necessary users, using 24-7/365 availability.

Benefits

The purpose of a contractor management program is to better centralize, qualify and monitor a contingent workforce. As a result of implementing a contractor management program, an organization can expect to experience some, or all, of the following advantages:

1. Cost savings
2. Better supplier/client relationships
3. Higher quality contractors and suppliers
4. Less paperwork for both owners and contractors. Some contractor management programs are cloud based software platforms that allow contractors to manage their own permits, licenses, inductions, training records and other issues that facilitate organisational and regulatory compliance.
5. Instant information sharing & evergreen qualification
6. Reduced risk – continuous improvement in loss control
7. Moving towards using leading indicators vs. lagging indicators
8. Contractor awareness with regulatory and best practices such as VPP and PSM

These advantages are both immediate and long-standing. A comprehensive web-based contractor management program from a reputable firm can provide a reliable basis for prequalifying contractors, vendors and other suppliers of goods and services.

Mitigating Risk

There are two major considerations when managing contractors. First is deciding on the criteria for evaluation and second is developing an effective management process to evaluate these criteria. There are a number of criteria on which a contractor's safety can be evaluated, such as historical and future trend information.

Contract Management Software

Contract management software is the range of computer programmes, libraries and data used to support contract management, contract lifecycle management, and contractor management on projects. It may be used with project management software.

Advantages and Key Functions

Most sophisticated projects involving contractors now use contract management software instead of relying on the manual management of paper contracts. It has become an essential tool for keeping track of multiple activities with cost implications, and can be especially helpful for automating administration, ensuring compliance and triggering alerts.

Contract management software is produced by many companies, working on a range of scales and offering varying degrees of customisability. Basic functions should include the ability to store contract documents, track changes to contract documents, search documents for a particular criterion, and to report required aspects of the contract. Other functions include managing a new contract request, capturing related data, following a document through a review and approval process, and collecting digital signatures.

Contract management software may also be an aid to project portfolio management and spend analysis, and may also monitor KPIs. Leading contract management software provides contract visibility, monitoring, and compliance to automate and streamline the contract lifecycle process.

Since a contract involves a client or operator working with one or more separate contractors, contract management software must assist all parties in communicating and keeping records across company boundaries.

A centralized repository provides a critical advantage allowing for all contract documents to be stored and managed within one location. Having contracts stored in multiple locations can delay and interrupt the contracting process.

Contract Risk Management Software (CRMS) for Capital Projects

Very large enterprises, such as capital expenditure (capex) projects, involve multiple parties and high risk and uncertainty. They are unlike traditional operating contracts in that they are subject to shared deadlines in unique situations. As the complexity of these unique projects increases, the relationships between parties become more important. This requires contract management software, or contract risk management software (CRMS), to become more dynamic and responsive.

The terms of these capex contracts necessarily involve assumptions at the start of the process, and are likely to change over the lifetime of the project lifecycle. For this reason, CRMS must be capable of recording one single instance of agreed changes to contract terms, and incorporating these changes in an auditable and legally robust way. With multiple decision makers involved, CRMS should also make accountability more transparent and enable faster decisions about variation proposals.

Disputes and Arbitration

During the performance of the contract, disputes may arise between the employer, engineer and contractor for various reasons of default in performance, progress payment, rates, time, etc. Negotiated amicable settlement is the endurable, quick solution, which requires mutual trust, cooperation, give and take policy, flexibility instead of rigidity of one's view, and vision for future relations. Unresolved disputes can conveniently be resolved through arbitration, which bears the stamps of law of natural justice, than resolution in court. Again, arbitration cannot succeed unless there is cooperation between contractor, client and arbitrator(s). Many times, when the award received is not palatable to either of the contesting parties, the unhappy party tries to drag on further in court to set aside the award. This defeats the very purpose of the arbitration process. The court does not set aside the award on flimsy ground or sits again for the judgment but on specific grounds. The court may even ask the arbitrator to review his award if any item is omitted. The responsibility imposed on the arbitrator is conduct the proceedings with prudence and diligence and not arbitrarily or whimsically and discharge his responsibilities in a cordial manner. The reasons for setting aside the award has been set out in Arbitration Act, 1940, which has since been modified in Arbitration & Conciliation Act 1996. As per the new Act, the award itself is considered as a decree and it need not be filed in the court. It is final and binding on both parties unless disputed by either party. The tendency to challenge the award should vanish, instead the aim should be to reach a negotiated, agreed, amicable and early settlement. It is hoped that better awareness and acceptance of this method fructifies in future.

Why Does Disputes Arise?

1. There are errors, ambiguities and omissions in the drawing and specification. Also, there is lack of proper coordination during construction.
2. Not complying with the intent of the contract or not adhering to the standards in the performance of work(Quality of conformance problem).
3. Incomplete, delayed, inaccurate response to the question by any of the party in the contract.
4. Unforeseen changes in sub surface conditions.
5. If site condition differs from those described in contract documents.
6. Extra work or changed work order
7. Not meeting schedule by the contractor.
8. Inadequate financial strength on the part of the owner, contractor or subcontractor.

Dispute Resolution

Dispute resolution is the process of resolving disputes between parties.

Methods

Methods of dispute resolution include:

- lawsuits (litigation)
- arbitration
- collaborative law
- mediation
- conciliation
- many types of negotiation
- facilitation

One could theoretically include violence or even war as part of this spectrum, but dispute resolution practitioners do not usually do so; violence rarely ends disputes effectively, and indeed, often only escalates them.

Dispute resolution processes fall into two major types:

1. Adjudicative processes, such as litigation or arbitration, in which a judge, jury or arbitrator determines the outcome.

2. Consensual processes, such as collaborative law, mediation, conciliation, or negotiation, in which the parties attempt to reach agreement.

Not all disputes, even those in which skilled intervention occurs, end in resolution. Such intractable disputes form a special area in dispute resolution studies.

Dispute resolution is an important requirement in international trade, including negotiation, mediation, arbitration and litigation.

Judicial Dispute Resolution

The legal system provides resolutions for many different types of disputes. Some disputants will not reach agreement through a collaborative process. Some disputes need the coercive power of the state to enforce a resolution. Perhaps more importantly, many people want a professional advocate when they become involved in a dispute, particularly if the dispute involves perceived legal rights, legal wrongdoing, or threat of legal action against them.

The most common form of judicial dispute resolution is litigation. Litigation is initiated when one party files suit against another. In the United States, litigation is facilitated by the government within federal, state, and municipal courts. The proceedings are very formal and are governed by rules, such as rules of evidence and procedure, which are established by the legislature. Outcomes are decided by an impartial judge and/or jury, based on the factual questions of the case and the application law. The verdict of the court is binding, not advisory; however, both parties have the right to appeal the judgment to a higher court. Judicial dispute resolution is typically adversarial in nature, for example, involving antagonistic parties or opposing interests seeking an outcome most favorable to their position.

Retired judges or private lawyers often become arbitrators or mediators; however, trained and qualified non-legal dispute resolution specialists form a growing body within the field of alternative dispute resolution (ADR). In the United States, many states now have mediation or other ADR programs annexed to the courts, to facilitate settlement of lawsuits.

Extrajudicial Dispute Resolution

Some use the term *dispute resolution* to refer only to alternative dispute resolution (ADR), that is, extrajudicial processes such as arbitration, collaborative law, and mediation used to resolve conflict and potential conflict between and among individuals, business entities, governmental agencies, and (in the public international law context) states. ADR generally depends on agreement by the parties to use ADR processes, either before or after a dispute has arisen. ADR has experienced steadily increasing acceptance and utilization because of a perception of greater flexibility, costs below those of traditional litigation, and speedy resolution of disputes, among other perceived advantages. However, some have criticized these methods as taking away the right to seek redress of grievances in the courts, suggesting that extrajudicial dispute resolution may not offer the fairest way for parties not in an equal bargaining relationship, for example in a dispute between a consumer and a large corporation. In addition, in some circumstances, arbitration and other ADR processes may become as expensive as litigation or more so.

Online Dispute Resolution

Dispute resolution can also take place on-line or by using technology in certain cases. Online dispute resolution, a growing field of dispute resolution, uses new technologies to solve disputes. Online Dispute Resolution is also called "ODR". Online Dispute Resolution or ODR also involves the application of traditional dispute resolution methods to disputes which arise online.

Arbitration

The London Court of International Arbitration.

Arbitration, a form of alternative dispute resolution (ADR), is a technique for the resolution of disputes outside the courts. The parties to a dispute refer it to arbitration by one or more persons (the "arbitrators", "arbiters" or "arbitral tribunal"), and agree to be bound by the arbitration decision (the "award"). A third party reviews the evidence in the case and imposes a decision that is legally binding on both sides and enforceable in the courts.

Arbitration is often used for the resolution of commercial disputes, particularly in the context of international commercial transactions. In certain countries such as the United States, arbitration is also frequently employed in consumer and employment matters, where arbitration may be mandated by the terms of employment or commercial contracts.

Arbitration can be either voluntary or mandatory (although mandatory arbitration can only come from a statute or from a contract that is voluntarily entered into, in which the parties agree to hold all existing or future disputes to arbitration, without necessarily knowing, specifically, what disputes will ever occur) and can be either binding or non-binding. Non-binding arbitration is similar to mediation in that a decision cannot be imposed on the parties. However, the principal distinction is that whereas a mediator will try to help the parties find a middle ground on which to compromise, the (non-binding) arbitrator remains totally removed from the settlement process and will only give a determination of liability and, if appropriate, an indication of the quantum of damages payable. By one definition arbitration is binding and non-binding arbitration is therefore technically not arbitration.

Arbitration is a proceeding in which a dispute is resolved by an impartial adjudicator whose decision the parties to the dispute have agreed, or legislation has decreed, will be final and binding. There are limited rights of review and appeal of arbitration awards. Arbitration is not the same as:

- judicial proceedings, although in some jurisdictions, court proceedings are sometimes referred as arbitrations
- alternative dispute resolution (ADR)
- expert determination
- mediation (a form of settlement negotiation facilitated by a neutral third party)

Advantages and Disadvantages

Parties often seek to resolve disputes through arbitration because of a number of perceived potential advantages over judicial proceedings:

- In contrast to litigation, where one cannot "choose the judge", arbitration allows the parties to choose their own tribunal. This is especially useful when the subject matter of the dispute is highly technical: arbitrators with an appropriate degree of expertise (for example, quantity surveying expertise, in the case of a construction dispute, or expertise in commercial property law, in the case of a real estate dispute) can be chosen.
- Arbitration is often faster than litigation in court.
- Arbitral proceedings and an arbitral award are generally non-public, and can be made confidential
- In arbitral proceedings the language of arbitration may be chosen, whereas in judicial proceedings the official language of the country of the competent court will be automatically applied.
- Because of the provisions of the New York Convention 1958, arbitration awards are generally easier to enforce in other nations than court verdicts.
- In most legal systems there are very limited avenues for appeal of an arbitral award, which is sometimes an advantage because it limits the duration of the dispute and any associated liability.

Some of the disadvantages include:

- Arbitration agreements are sometimes contained in ancillary agreements, or in small print in other agreements, and consumers and employees often do not know in advance that they have agreed to mandatory binding pre-dispute arbitration by purchasing a product or taking a job.

- If the arbitration is mandatory and binding, the parties waive their rights to access the courts and to have a judge or jury decide the case.
- If the arbitrator or the arbitration forum depends on the corporation for repeat business, there may be an inherent incentive to rule against the consumer or employee
- There are very limited avenues for appeal, which means that an erroneous decision cannot be easily overturned.
- Although usually thought to be speedier, when there are multiple arbitrators on the panel, juggling their schedules for hearing dates in long cases can lead to delays.
- In some legal systems, arbitration awards have fewer enforcement options than judgments; although in the United States arbitration awards are enforced in the same manner as court judgments and have the same effect.
- Arbitrators are generally unable to enforce interlocutory measures against a party, making it easier for a party to take steps to avoid enforcement of member or a small group of members in arbitration due to increasing legal fees, without explaining to the members the adverse consequences of an unfavorable ruling.
- Discovery may be more limited in arbitration or entirely nonexistent.
- The potential to generate billings by attorneys may be less than pursuing the dispute through trial.
- Unlike court judgments, arbitration awards themselves are not directly enforceable. A party seeking to enforce an arbitration award must resort to judicial remedies, called an action to "confirm" an award.

Arbitrability

By their nature, the subject matter of some disputes is not capable of arbitration. In general, two groups of legal procedures cannot be subjected to arbitration:

- Procedures which necessarily lead to a determination which the parties to the dispute may not enter into an agreement upon: Some court procedures lead to judgments which bind all members of the general public, or public authorities in their capacity as such, or third parties, or which are being conducted in the public interest. For example, until the 1980s, antitrust matters were not arbitrable in the United States. Matters relating to crimes, status and family law are generally not considered to be arbitrable, as the power of the parties to enter into an agreement upon these matters is at least restricted. However, most other disputes that involve private rights between two parties can be resolved using arbitration. In some disputes, parts of claims may be arbitrable and other parts not. For example, in a dispute over patent infringement, a determination of whether a patent has been infringed could be adjudicated upon by an arbitration tribunal, but the validity of a patent could not: As patents are subject to a system of public registration, an arbitral panel would have no power to order the relevant body to rectify any patent registration based upon its determination.

- Some legal orders exclude or restrict the possibility of arbitration for reasons of the protection of weaker members of the public, e.g. consumers. *Examples*: German law excludes disputes over the rental of living space from any form of arbitration, while arbitration agreements with consumers are only considered valid if they are signed by either party, and if the signed document does not bear any other content than the arbitration agreement.

Arbitration Agreement

Arbitration agreements are generally divided into two types:

- Agreements which provide that, if a dispute should arise, it will be resolved by arbitration. These will generally be normal contracts, but they contain an arbitration clause
- Agreements which are signed after a dispute has arisen, agreeing that the dispute should be resolved by arbitration (sometimes called a "submission agreement")

The former is the far more prevalent type of arbitration agreement. Sometimes, legal significance attaches to the type of arbitration agreement. For example, in certain Commonwealth countries (not including England and Wales), it is possible to provide that each party should bear their own costs in a conventional arbitration clause, but not in a submission agreement.

In keeping with the informality of the arbitration process, the law is generally keen to uphold the validity of arbitration clauses even when they lack the normal formal language associated with legal contracts. Clauses which have been upheld include:

- "arbitration in London - English law to apply"
- "suitable arbitration clause"
- "arbitration, if any, by ICC Rules in London"

The courts have also upheld clauses which specify resolution of disputes other than in accordance with a specific legal system. These include provision indicating:

- That the arbitrators "must not necessarily judge according to the strict law but as a general rule ought chiefly to consider the principles of practical business"
- "internationally accepted principles of law governing contractual relations"

Agreements to refer disputes to arbitration generally have a special status in the eyes of the law. For example, in disputes on a contract, a common defence is to plead the contract is void and thus any claim based upon it fails. It follows that if a party successfully claims that a contract is void, then each clause contained within the contract, including the arbitration clause, would be void. However, in most countries, the courts have accepted that:

1. A contract can only be declared void by a court or other tribunal; and
2. If the contract (valid or otherwise) contains an arbitration clause, then the proper forum to determine whether the contract is void or not, is the arbitration tribunal.

Arguably, either position is potentially unfair; if a person is made to sign a contract under duress, and the contract contains an arbitration clause highly favourable to the other party, the dispute may still referred to that arbitration tribunal. Conversely a court may be persuaded that the arbitration agreement itself is void having been signed under duress. However, most courts will be reluctant to interfere with the general rule which does allow for commercial expediency; any other solution (where one first had to go to court to decide whether one had to go to arbitration) would be self-defeating.

Comparative Law

Nations regulate arbitration through a variety of laws. The main body of law applicable to arbitration is normally contained either in the national Private International Law Act (as is the case in Switzerland) or in a separate law on arbitration (as is the case in England and Jordan). In addition to this, a number of national procedural laws may also contain provisions relating to arbitration.

United States

The Federal Arbitration Act (FAA) of 1925 established a public policy in favor of arbitration. For the first six decades of its existence, courts did not allow arbitration for "federal statutory claims" through a bright-line "nonarbitrability" doctrine, but in the 1980s the Supreme Court of the United States reversed and began to use the act to require arbitration if included in the contract for federal statutory claims. Although some legal scholars believe that it was originally intended to apply to federal courts only, courts now routinely require arbitration due to the FAA regardless of state statutes or public policy unconscionability determinations by state courts. In consumer law, standard form contracts often include mandatory predispute arbitration clauses which require consumer arbitration. Under these agreements the consumer may waive their right to a lawsuit and a class action. In 2011, one of these clauses was upheld in *AT&T Mobility v. Concepcion.*

Several arbitration organizations exist, including the American Arbitration Association and JAMS. The National Arbitration Forum also conducts arbitrations, but it no longer conducts consumer arbitrations pursuant to a consent decree entered into in 2009 because of evidence that it had been biased toward, and had incentives that favored, credit card companies over cardholders. The AAA was also asked to exit the business., but has not done so.

International

History

The United States and Great Britain were pioneers in the use of arbitration to resolve their differences. It was first used in the Jay Treaty of 1795, and played a major role in the Alabama Claims case of 1872 whereby major tensions regarding British support for the Confederacy during the American Civil War were resolved. At the First International Conference of American States in 1890, a plan for systematic arbitration was developed, but not excepted. The Hague Peace Conference of 1899, saw the major world powers agreed to a system of arbitration and the creation of a Permanent Court of Arbitration. President William Howard Taft was a major advocate. One important use came in the Newfoundland fisheries dispute between the United States and Britain in 1910. In 1911 the United States signed arbitration treaties with France and Britain.

Arbitration was widely discussed among diplomats and elites in the 1890-1914 era. The 1895 dispute between the United States and Britain over Venezuela was peacefully resolved through arbitration. Both nations realized that a mechanism was desirable to avoid possible future conflicts. The Olney-Pauncefote Treaty of 1897 was a proposed treaty between the United States and Britain in 1897 that required arbitration of major disputes. The treaty was rejected by the U.S. Senate and never went into effect.

American Secretary of State William Jennings Bryan (1913–1915) worked energetically to promote international arbitration agreements, but his efforts were frustrated by the outbreak of World War I. Bryan negotiated 28 treaties that promised arbitration of disputes before war broke out between the signatory countries and the United States. He made several attempts to negotiate a treaty with Germany, but ultimately was never able to succeed. The agreements, known officially as "Treaties for the Advancement of Peace," set up procedures for conciliation rather than for arbitration. Arbitration treaties were negotiated after the war, but attracted much less attention than the negotiation mechanism created by the League of Nations.

International Agreements

By far the most important international instrument on arbitration law is the 1958 New York Convention on Recognition and Enforcement of Foreign Arbitral Awards, usually simply referred to as the "New York Convention". Virtually every significant commercial country is a signatory, and only a handful of countries are not parties to the New York Convention.

Some other relevant international instruments are:

- The Geneva Protocol of 1923
- The Geneva Convention of 1927
- The European Convention of 1961
- The Washington Convention of 1965 (governing settlement of international investment disputes)
- The Washington Convention (ICSID) of 1996 for investment arbitration
- The UNCITRAL Model Law on International Commercial Arbitration of 1985, (revised in 2006).
- The UNCITRAL Arbitration Rules (providing a set of rules for an ad hoc arbitration)

International Enforcement

It is often easier to enforce arbitration awards in a foreign country than court judgments. Under the New York Convention 1958, an award issued in a contracting state can generally be freely enforced in any other contracting state, only subject to certain, limited defenses. Only foreign arbitration awards are enforced pursuant to the New York Convention. An arbitral decision is foreign where the award was made in a state other than the state of recognition or where foreign procedural law was used. In most cases, these disputes are settled with no public record of their existence as the loser complies voluntarily, although in 2014 UNCITRAL promulgated a rule for public disclosure of investor-state disputes.

Virtually every significant commercial country in the world is a party to the Convention while relatively few countries have a comprehensive network for cross-border enforcement of judgments their courts. Additionally, the awards not limited to damages. Whereas typically only monetary judgments by national courts are enforceable in the cross-border context, it is theoretically possible (although unusual in practice) to obtain an enforceable order for specific performance in an arbitration proceeding under the New York Convention.

Article V of the New York Convention provides an exhaustive list of grounds on which enforcement can be challenged. These are generally narrowly construed to uphold the pro-enforcement bias of the Convention.

Government Disputes

Certain international conventions exist in relation to the enforcement of awards against states.

- The Washington Convention 1965 relates to settlement of investment disputes between states and citizens of other countries. The Convention created the International Centre for Settlement of Investment Disputes (or ICSID). Compared to other arbitration institutions, relatively few awards have been rendered under ICSID.
- The Algiers Declaration of 1981 established the Iran-US Claims Tribunal to adjudicate claims of American corporations and individuals in relation to expropriated property during the Islamic revolution in Iran in 1979. The tribunal has not been a notable success, and has even been held by an English court to be void under its own governing law.

Arbitral Tribunal

The arbitrators which determine the outcome of the dispute are called the arbitral tribunal. The composition of the arbitral tribunal can vary enormously, with either a sole arbitrator sitting, two or more arbitrators, with or without a chairman or umpire, and various other combinations. In most jurisdictions, an arbitrator enjoys immunity from liability for anything done or omitted whilst acting as arbitrator unless the arbitrator acts in bad faith.

Arbitrations are usually divided into two types: *ad hoc* arbitrations and administered arbitrations.

In *ad hoc* arbitrations, the arbitral tribunals are appointed by the parties or by an appointing authority chosen by the parties. After the tribunal has been formed, the appointing authority will normally have no other role and the arbitation will be managed by the tribunal.

In administered arbitration, the arbitration will be administered by a professional arbitration institution providing arbitration services, such as the LCIA in London, or the ICC in Paris, or the American Arbitration Association in the United States. Normally the arbitration institution also will be the appointing authority. Arbitration institutions tend to have their own rules and procedures, and may be more formal. They also tend to be more expensive, and, for procedural reasons, slower.

Duties of the Tribunal

The duties of a tribunal will be determined by a combination of the provisions of the arbitration agreement and by the procedural laws which apply in the seat of the arbitration. The extent to which the laws of the seat of the arbitration permit "party autonomy" (the ability of the parties to set out their own procedures and regulations) determines the interplay between the two.

However, in almost all countries the tribunal owes several non-derogable duties. These will normally be:

- to act fairly and impartially between the parties, and to allow each party a reasonable opportunity to put their case and to deal with the case of their opponent (sometimes shortened to: complying with the rules of "natural justice"); and
- to adopt procedures suitable to the circumstances of the particular case, so as to provide a fair means for resolution of the dispute.

Arbitral Awards

Although arbitration awards are characteristically an award of damages against a party, in many jurisdictions tribunals have a range of remedies that can form a part of the award. These may include:

1. payment of a sum of money (conventional damages)
2. the making of a "declaration" as to any matter to be determined in the proceedings
3. in some jurisdictions, the tribunal may have the same power as a court to:
    1. order a party to do or refrain from doing something ("injunctive relief")
    2. to order specific performance of a contract
    3. to order the rectification, setting aside or cancellation of a deed or other document.
4. In other jurisdictions, however, unless the parties have expressly granted the arbitrators the right to decide such matters, the tribunal's powers may be limited to deciding whether a party is entitled to damages. It may not have the legal authority to order injunctive relief, issue a declaration, or rectify a contract, such powers being reserved to the exclusive jurisdiction of the courts.

Challenge

Generally speaking, by their nature, arbitration proceedings tend not to be subject to appeal, in the ordinary sense of the word. However, in most countries, the court maintains a supervisory role to set aside awards in extreme cases, such as fraud or in the case of some serious legal irregularity on the part of the tribunal. Only domestic arbitral awards are subject to set aside procedure.

In American arbitration law there exists a small but significant body of case law which deals with the power of the courts to intervene where the decision of an arbitrator is in fundamental disaccord with the applicable principles of law or the contract. However, this body of case law has been called into question by recent decisions of the Supreme Court.

Unfortunately there is little agreement amongst the different American judgments and textbooks as to whether such a separate doctrine exists at all, or the circumstances in which it would apply. There does not appear to be any recorded judicial decision in which it has been applied. However, conceptually, to the extent it exists, the doctrine would be an important derogation from the general principle that awards are not subject to review by the courts.

Costs

The overall costs of arbitration can be estimated on the websites of international arbitration institutions, such as that of the ICC, the website of the SIAC and the website of the International Arbitration Attorney Network. The overall cost of administrative and arbitrator fees is, on average, less than 20% of the total cost of international arbitration.

In many legal systems - both common law and civil law - it is normal practice for the courts to award legal costs against a losing party, with the winner becoming entitled to recover an approximation of what it spent in pursuing its claim (or in defense of a claim). The United States is a notable exception to this rule, as except for certain extreme cases, a prevailing party in a US legal proceeding does not become entitled to recoup its legal fees from the losing party.

Like the courts, arbitral tribunals generally have the same power to award costs in relation to the determination of the dispute. In international arbitration as well as domestic arbitrations governed by the laws of countries in which courts may award costs against a losing party, the arbitral tribunal will also determine the portion of the arbitrators' fees that the losing party is required to bear.

Nomenclature

As methods of dispute resolution, arbitration procedure can be varied to suit the needs of the parties. Certain specific "types" of arbitration procedure have developed, particularly in North America.

- Judicial Arbitration is, usually, not arbitration at all, but merely a court process which refers to itself as arbitration, such as small claims arbitration before the County Courts in the United Kingdom.
- Online Arbitration is, a form of arbitration that occurs exclusively online. There is currently an assumption that online arbitration is admissible under the New York Convention and the E-Commerce Directive, but this has not been legally verified. Since arbitration is based on a contractual agreement between the parties, an online process without a regulatory framework may generate a significant number of challenges from consumers and other weaker parties if due process cannot be assured.
- High-Low Arbitration, or Bracketed Arbitration, is an arbitration wherein the parties to the dispute agree in advance the limits within which the arbitral tribunal must render its award. It is only generally useful where liability is not in dispute, and the only issue be-

tween the party is the amount of compensation. If the award is lower than the agreed minimum, then the defendant only need pay the lower limit; if the award is higher than the agreed maximum, the claimant will receive the upper limit. If the award falls within the agreed range, then the parties are bound by the actual award amount. Practice varies as to whether the figures may or may not be revealed to the tribunal, or whether the tribunal is even advised of the parties' agreement.

- Binding Arbitration is a form of arbitration where the decision by the arbitrator is legally binding and enforceable, similar to a court order.
- Non-Binding Arbitration is a process which is conducted as if it were a conventional arbitration, except that the award issued by the tribunal is not binding on the parties, and they retain their rights to bring a claim before the courts or other arbitration tribunal; the award is in the form of an independent assessment of the merits of the case, designated to facilitate an out-of-court settlement. State law may automatically make a non-binding arbitration binding, if, for example, the non-binding arbitration is court-ordered, and no party requests a trial de novo (as if the arbitration had not been held).
- Pendulum Arbitration refers to a determination in industrial disputes where an arbitrator has to resolve a claim between a trade union and management by making a determination of which of the two sides has the more reasonable position. The arbitrator must choose only between the two options, and cannot split the difference or select an alternative position. It was initiated in Chile in 1979. This form of arbitration has been increasingly seen in resolving international tax disputes, especially in the context of deciding on the Transfer Pricing margins. This form of arbitration is also known (particularly in the United States) as Baseball Arbitration. It takes its name from a practice which arose in relation to salary arbitration in Major League Baseball.
- Night Baseball Arbitration is a variation of baseball arbitration where the figures are not revealed to the arbitration tribunal. The arbitrator will determinate the quantum of the claim in the usual way, and the parties agree to accept and be bound by the figure which is closest to the tribunal's award.

Such forms of "Last Offer Arbitration" can also be combined with mediation to create MEDALOA hybrid processes (Mediation followed by Last Offer Arbitration).

History

England

Arbitration in its common law form developed in England; in the Middle Ages, tribunals such as the Courts of the Boroughs, of the Fair and of the Staple arose as the Royal Courts were not designed for trade disputes, and trade with foreigners was otherwise unenforceable. In the mid-16th century, common law courts developed contract law and the Admiralty court became accessible for disputes with foreign merchants, broadening the venues for trade disputes. Courts became suspicious of arbitration; for example, in *Kill v. Hollister* (1746), an English court ruled that the arbitration agreement could 'oust' courts of law and equity of jurisdiction. Merchants, however, retained provisions to settle disputes among themselves, but tension between the arbitration proceedings and courts eventually resulted in the Common Law Procedure Act 1854 which provided for the appointment of arbitrators and umpires, allowed courts to 'stay proceedings' when a disputant filed a suit despite an agreement to arbitrate, and provided a process for arbitrators to submit questions to a court. Later, the Arbitration Act 1889 was passed, followed by other Arbitration Acts in 1950, 1975, 1979 and 1996. Arbitration Act 1979 in particular limited judicial review for arbitration awards.

United States

Arbitration was common in the early United States, with George Washington serving as an arbiter on an occasion. The United States had a notable difference from England, however, in that unlike England, its courts generally did not enforce executory agreements (binding predispute agreements) to arbitrate. This meant that prior to an award, a claimant could sue in court even if they had contractually agreed to settle disputes by arbitration. After the award, courts reviewed the judgment, but generally deferred to the arbitration, although the practice was not consistent.

The lack of enforcement of predispose agreements led to the Federal Arbitration Act of 1925, with New York leading with a state law enforcing predispose agreements. In 1921, the American Bar Association drafted the Federal Arbitration Act based on the New York law, which was passed in 1925 with minor changes. In the next decade, the American Arbitration Association promoted rules and facilitated arbitrations through appointments.

Conciliation

Conciliation is an alternative dispute resolution (ADR) process whereby the parties to a dispute use a conciliator, who meets with the parties both separately and together in an attempt to resolve their differences. They do this by lowering tensions, improving communications, interpreting issues, encouraging parties to explore potential solutions and assisting parties in finding a mutually acceptable outcome.

Conciliation differs from arbitration in that the conciliation process, in and of itself, has no legal standing, and the conciliator usually has no authority to seek evidence or call witnesses, usually writes no decision, and makes no award.

Conciliation differs from mediation in that in conciliation, often the parties are in need of restoring or repairing a relationship, either personal or business.

Effectiveness

Recent studies in the processes of negotiation have indicated the effectiveness of a technique that deserves mention here. A conciliator assists each of the parties to independently develop a list of all of their objectives (the outcomes which they desire to obtain from the conciliation). The conciliator then has each of the parties separately prioritize their own list from most to least important. He/She then goes back and forth between the parties and encourages them to "give" on the objectives one at a time, starting with the least important and working toward the most important for each party in turn. The parties rarely place the same priorities on all objectives, and usually have some objectives that are not listed by the other party. Thus the conciliator can quickly build a string of successes and help the parties create an atmosphere of trust which the conciliator can continue to develop.

Most successful conciliators are highly skilled negotiators. Some conciliators operate under the auspices of any one of several non-governmental entities, and for governmental agencies such as the Federal Mediation and Conciliation Service in the United States.

Historical Conciliation

Historical conciliation is an applied conflict resolution approach that utilizes historical narratives to positively transform relations between societies in conflicts. Historical conciliation can utilize many different methodologies, including mediation, sustained dialogue, apologies, acknowledgement, support of public commemoration activities, and public diplomacy.

Historical conciliation is not an excavation of objective facts. The point of facilitating historical questions is not to discover all the facts in regard to who was right or wrong. Rather, the objective is to discover the complexity, ambiguity, and emotions surrounding both dominant and non-dominant cultural and individual narratives of history. It is also not a rewriting of history. The goal is not to create a combined narrative that everyone agrees upon. Instead, the aim is to create room for critical thinking and more inclusive understanding of the past and conceptions of "the other."

Conflicts that are addressed through historical conciliation have their roots in conflicting identities of the people involved. Whether the identity at stake is their ethnicity, religion or culture, it requires a comprehensive approach that takes people's needs, hopes, fears, and concerns into account.

Japan

Japanese law makes extensive use of conciliation in civil disputes. The most common forms are civil conciliation and domestic conciliation, both of which are managed under the auspice of the court system by one judge and two non-judge "conciliators."

Civil conciliation is a form of dispute resolution for small lawsuits, and provides a simpler and cheaper alternative to litigation. Depending on the nature of the case, non-judge experts (doctors, appraisers, actuaries, and so on) may be called by the court as conciliators to help decide the case.

Domestic conciliation is most commonly used to handle contentious divorces, but may apply to other domestic disputes such as the annulment of a marriage or acknowledgment of paternity. Parties in such cases are required to undergo conciliation proceedings and may only bring their case to court once conciliation has failed.

References

- Eriksson, Erik C.; Westerberg, Mats (2011). "Effects of cooperative procurement procedures on construction project performance: A conceptual framework". International Journal of Project Management. 29 (2): 197–208. doi:10.1016/j.ijproman.2010.01.003
- Vitasek, Kate, et. al. (2011). The Vested Outsourcing Manual (1st ed.). New York: Palgrave Macmillan. ISBN 0230112684
- Li, Huimin C.; Arditi, David; Wang, Zhuofu (2013). "Factors That Affect Transaction Costs in Construction Projects". Journal of Construction Engineering and Management. 139 (1): 60–68. doi:10.1061/(ASCE)CO.1943-7862.0000573

- "Best Practices in Contract Management: Strategies for Optimizing Business Relationships<". Aberdeen Group. Retrieved 2008-07-10
- Contract & Commercial Management - The Operational Guide, Van Haren Publishing, October 2011, ISBN 978 90 8753 627 5
- Cardenas, IC; Voordijk, H; Dewulf, G (2017). "Beyond theory: Towards a probabilistic causation model to support project governance in infrastructure projects". International Journal of Project Management. doi:10.1016/j.ijproman.2017.01.002
- "Contract Management: Optimizing Revenues and Capturing Savings". Aberdeen Group. May 2007. Retrieved 2008-07-10
- Chen, Le; Karen, Manley (2014). "Validation of an instrument to measure governance and performance on collaborative infrastructure projects". Journal of Construction Engineering and Management. 140 (5). doi:10.1061/(ASCE)CO.1943-7862.0000834

3

Estimation and Rate Analysis

For any form of construction work, it is necessary to determine the cost to carry out and complete the project. The construction project manager estimates the quantity and cost. This is known as the estimate. Construction work estimate is classified into preliminary and detailed estimate. The topics discussed in the section are of great importance to broaden the existing knowledge on estimation concepts and rate analysis.

Estimation

During construction project planning and implementation, we need to know the quantities and costs of various items required to meet the objective of the project. That is, construction project manager has to anticipate the cost of project. The process of calculation of quantities and costs of various items in connection with the construction project is called an "estimate".

An estimate is prepared by first obtaining the quantity of the items required to complete the project and multiplied by unit cost of the items. Details of the estimate depend upon the purpose of carrying out estimate.

The preliminary estimate should be accurate enough to produce the cost of the project within 10% to 15% variations.

During the life cycle of a project, different parties for various reasons require an estimate. The details and method of estimations depend upon the needs and level of management decisions. Suppose a philanthropic organization wants to start an academic institute. The organization is interested in knowing how much money is required to be spent. It can be just calculated from the experience with previous project of similar kind. Such as, 20 thousand money per student has to be invested. If academic institute is to be planned for one thousand students, 20 x 1000 thousand is required. Next level of details may be required is how much money is to be invested on different subheads, equipment, laboratories, hostels etc.

Suppose a philanthropic organization wants to start an academic institute. The organization is interested in knowing how much money is required to be spent. It can be just calculated from the experience with previous project of similar kind. Such as, 20 thousand money per student has to be invested. If academic institute is to be planned for one thousand students, 20 x 1000 thousand is required. Next level of details may be required is how much money is to be invested on different subheads, equipment, laboratories, hostels etc. These can be calculated based on knowing the area of construction and cost of construction per unit area. But suppose hostel has to be built based on specific drawing and specification. Now an estimate will be prepared from the drawing and specification using specified rules. This estimates form the basis for contract evaluation by the client.

Similarly when contractor is bidding for the hostel, he will prepare his own estimate such that he can buy the contract, also gets profit from the execution of the project. But during construction period the construction, company has to prepare an estimate for order and supply of the material, also for resource planning.

Thus we have seen from the above example that the estimate of same item of works differs in details based as need of the project implementation. Also the level of accuracy needed for the estimate is dependent on the purpose of estimate. Like estimate for material supply has to be very accurate because storage of material will be the burden for the project. But the accuracy of the estimate to know the total cost to study the feasibility of the project will not be very high.

Different body involved in preparation of estimate can be grouped into following three classes.

1. The owner or owner's staff such as project manager.
2. Contractor's personnel such as senior manager, planner, site management staffs etc.
3. Materials suppliers and contractors

The owner generally employs a project manager/engineer, who helps him in preparation of contract documents. The contract documents require the bill of quantities. Also, owner should be aware of the cost to sell the contract. Thus an estimate is required of various quantities. The owner's project manager prepares these estimates. Tender for the contract in a construction company is prepared by higher management. These management are helped by estimators, site engineers and material supplier to prepare the estimate of the job.

The exact number of candies in this jar cannot be determined by looking at it, because most of the candies are not visible. The amount can be estimated by presuming that the portion of the jar that cannot be seen contains an amount equivalent to the amount contained in the same volume for the portion that can be seen.

Estimation (or estimating) is the process of finding an estimate, or approximation, which is a value that is usable for some purpose even if input data may be incomplete, uncertain, or unstable. The value is nonetheless usable because it is derived from the best information available. Typically, estimation involves "using the value of a statistic derived from a sample to estimate the value of a corresponding population parameter". The sample provides information that can be projected, through various formal or informal processes, to determine a range most likely to describe the missing information. An estimate that turns out to be incorrect will be an overestimate if the estimate exceeded the actual result, and an underestimate if the estimate fell short of the actual result.

How Estimation is Done

Estimation is often done by sampling, which is counting a small number of examples something, and projecting that number onto a larger population. An example of estimation would be determining how many candies of a given size are in a glass jar. Because the distribution of candies inside the jar may vary, the observer can count the number of candies visible through the glass, consider the size of the jar, and presume that a similar distribution can be found in the parts that can not be seen, thereby making an estimate of the total number of candies that could be in the jar if that presumption were true. Estimates can similarly be generated by projecting results from polls or surveys onto the entire population.

In making an estimate, the goal is often most useful to generate a range of possible outcomes that is precise enough to be useful, but not so precise that it is likely to be inaccurate. For example, in trying to guess the number of candies in the jar, if fifty were visible, and the total volume of the jar seemed to be about twenty times as large as the volume containing the visible candies, then one might simply project that there were a thousand candies in the jar. Such a projection, intended to pick the single value that is believed to be closest to the actual value, is called a point estimate. However, a point estimation is likely to be incorrect, because the sample size - in this case, the number of candies that are visible - is too small a number to be sure that it does not contain anomalies that differ from the population as a whole. A corresponding concept is an interval estimate, which captures a much larger range of possibilities, but is too broad to be useful. For example, if one were asked to estimate the percentage of people who like candy, it would clearly be correct that the number falls between zero and one hundred percent. Such an estimate would provide no guidance, however, to somebody who is trying to determine how many candies to buy for a party to be attended by a hundred people.

Uses of Estimation

In mathematics, approximation describes the process of finding estimates in the form of upper or lower bounds for a quantity that cannot readily be evaluated precisely, and approximation theory deals with finding simpler functions that are close to some complicated function and that can provide useful estimates. In statistics, an estimator is the formal name for the rule by which an estimate is calculated from data, and estimation theory deals with finding estimates with good properties. This process is used in signal processing, for approximating an unobserved signal on the basis of an observed signal containing noise. For estimation of yet-to-be observed quantities, forecasting and prediction are applied. A Fermi problem, in physics, is one concerning estimation in problems which typically involve making justified guesses about quantities that seem impossible to compute given limited available information.

Estimation is important in business and economics, because too many variables exist to figure out how large-scale activities will develop. Estimation in project planning can be particularly significant, because plans for the distribution of labor and for purchases of raw materials must be made, despite the inability to know every possible problem that may come up. A certain amount of resources will be available for carrying out a particular project, making it important to obtain or generate a cost estimate as one of the vital elements of entering into the project. The U.S. Government Accountability Office defines a cost estimate as, “the summation of individual cost elements, using established methods and valid data, to estimate the future costs of a program, based on what is known today”, and reports that “realistic cost estimating was imperative when making wise decisions in acquiring new systems”. Furthermore, project plans must not underestimate the needs of the project, which can result in delays while unmet needs are fulfilled, nor must they greatly overestimate the needs of the project, or else the unneeded resources may go to waste.

An informal estimate when little information is available is called a guesstimate, because the inquiry becomes closer to purely guessing the answer. The “estimated” sign, ℮, is used to designate that package contents are close to the nominal contents.

Types of Estimate

Based on the rigor with which an estimate is prepared, the estimate can be classified as

1. Preliminary estimate.
2. Detailed estimate.

Requirements of Preliminary and Detailed Estimate:

To prepare a preliminary estimate, quantity of work to be planned is measured in terms of unit of person, built up area, extent of work in kilometer such as a preliminary estimate of housing can be calculated based on type of house and number of house to be built. Similar preliminary estimate of highway can be calculated by knowing the type of road in terms of lane or pavement and length of the load.

Preliminary Estimate

Different methods of preliminary estimate can be grouped into the following categories –

- Time referenced cost estimate
- Cost capacity factor
- Component ration/installation cost ratio
- Parameter cost

The preliminary estimate is obtained during early phases of project. As the project progresses, an estimate keeps on getting refined and lastly we get definitive estimate.

i. Time referenced cost estimate – If the cost of construction of facilities is known at a certain time, the present cost can be estimated by multiplying it with the cost index, with reference to that time. Generally government finance department publishes cost index based on depreciation, etc. But this cost index is very crude. Construction department keeps the cost index based on the items and method of construction used. There are some limitations of such method. During the time, technology changes; productivity of various equipment may change; degree of competition may be different. One should be aware of all these factors while using the cost index to estimate the constructed facilities.

ii. Cost capacity factor – This factor refers to the change in cost of built-up facilities with the change in size of the project of similar type. If C1 and C2 be the cost of facilities to be built and old facilities. Let their size be Q1 and Q2 . We can calculate the cost of the new facility as

$$C_1 = C_2\left(\frac{Q_1}{Q_2}\right)^x$$

Where x is called the capacity factor. Many organizations publish the value of capacity factor. It is reasonable to consider cost capacity factor between 0.6 to 0.8. The capacity factor method and time reference method can be combined.

iii. Installation cost ratio – In the industrial project, the equipment to be installed is known quite earlier to the details for their installation. In such situations, an estimate is prepared based on the ratio or percentage of the cost incurred as the installation cost. The organization generally maintains this type of data. Example- The cost of an equipment is Rs10,00,000, and its installation cost , including accessories is 50%. Thus the estimated cost is

10,00,000 + (0.5 x 10,00,000) = 15,00,000 We can consider the equipment in component wise and calculate the installation cost individually. This approach will require detailed information, but will be more accurate.

iv. Unit Parameter cost – In many of the projects, it is possible to obtain the cost of unit parameter of construction. These parameters may be unit area, unit volume, etc. This type or procedures are used in building construction. Residential apartment in our country are generally estimated using this method. There are different variations of the method. Some of them widely used are as follows - Residential buildings are generally estimated using this method.

Requirement for Detailed Estimate:

To calculate the detailed estimate, complete structural drawing of the facilities to be built up and specification of the items are necessary. We also need to know the rate of each of the items, which includes profit of the contractor.

Revised Estimate:

Due to change in the rate of material or change in specification of materials, an estimate is often required to be modified. Thus, we prepare revised estimate against the original estimate. In the preparation of revised estimate, original estimate is also provided such that comparison can be made. There is no change in structural drawing and additional items are not added in the revised estimate.

Revised estimate is a detailed estimate. It is also accompanied by the original estimate for comparative purpose. While submitting the revised estimate, the reason for revision should also be stated.

Supplementary Estimate

During the execution of project, structural modification are required or addition work is added for construction. The estimate for additional work is called supplementary estimate. Supplementary estimate is also required to be prepared when some of the items are overlooked. In preparing the supplementary estimate reason for preparing the estimate would be clearly stated.

Example: Additional partition can be planned in the building. The estimate is prepared for partitioning as supplementary estimate. Structural / architectural changes are prerequisite for supplementary estimate.

Challenges in Estimating

The job of an estimator is not only to calculate the quantity and cost of the material, but to co-ordinate different parties involved in the project or construction in particular. An estimator of the construction company has to get involved with the site engineer, material supplier, instrument (plant) manager and senior management staff to obtain the various details of the cost and explaining the calculation to higher management. An estimator has to understand the project details, study drawing and specification, obtain working conditions, labour and equipment productivity and collecting other data from the site. An estimate also assesses the temporary work at the site such as formwork for concreting, temporary plant for the storage of materials. Virtually, an estimator has to build the project on paper to obtain the right guess.

In the organization, background work for estimation is carried out by higher management and organizations keeps the details regarding unit cost of the items called schedule of rate, productivity, and guideline for basic calculation. The basic calculation is carried out by junior level engineers or assistants.

In preparing detailed estimate, we need to know the following two things

1. Quantities of the items involved.
2. Obtaining cost of the items involved.

Methods of obtaining quantities of the items involved is called quantity estimate and method obtaining unit cost of the items involved is called rate analysis.

Cost

In production, research, retail, and accounting, a cost is the value of money that has been used up to produce something, and hence is not available for use anymore. In business, the cost may be one of acquisition, in which case the amount of money expended to acquire it is counted as cost. In this case, money is the input that is gone in order to acquire the thing. This acquisition cost may be the sum of the cost of production as incurred by the original producer, and further costs of transaction as incurred by the acquirer over and above the price paid to the producer. Usually, the price also includes a mark-up for profit over the cost of production.

More generalized in the field of economics, cost is a metric that is totaling up as a result of a process or as a differential for the result of a decision. Hence cost is the metric used in the standard modeling paradigm applied to economic processes.

Costs (pl.) are often further described based on their timing or their applicability.

Types of Accounting Costs

In accounting, costs are the monetary value of expenditures for supplies, services, labor, products, equipment and other items purchased for use by a business or other accounting entity. It is the amount denoted on invoices as the price and recorded in bookkeeping records as an expense or asset cost basis.

Opportunity cost, also referred to as *economic cost* is the value of the best alternative that was not chosen in order to pursue the current endeavor—i.e., what could have been accomplished with the resources expended in the undertaking. It represents opportunities forgone.

In theoretical economics, cost used without qualification often means opportunity cost.

Comparing Private, External, and Social Costs

When a transaction takes place, it typically involves both private costs and external costs.

Private costs are the costs that the buyer of a good or service pays the seller. This can also be described as the costs internal to the firm's production function.

External costs (also called externalities), in contrast, are the costs that people other than the buyer are forced to pay as a result of the transaction. The bearers of such costs can be either particular individuals or society at large. Note that external costs are often both non-monetary and problematic to quantify for comparison with monetary values. They include things like pollution, things that society will likely have to pay for in some way or at some time in the future, but that are not included in transaction prices.

Social costs are the sum of private costs and external costs.

For example, the manufacturing cost of a car (i.e., the costs of buying inputs, land tax rates for the car plant, overhead costs of running the plant and labor costs) reflects the *private cost* for the manufacturer (in some ways, normal profit can also be seen as a cost of production; see, e.g., Ison and Wall, 2007, p. 181). The polluted waters or polluted air also created as part of the process of producing the car is an *external cost* borne by those who are affected by the pollution or who value unpolluted air or water. Because the manufacturer does not pay for this external cost (the cost of emitting undesirable waste into the commons), and does not include this cost in the price of the car (a Kaldor-Hicks compensation), they are said to be external to the market pricing mechanism. The air pollution from driving the car is also an externality produced by the car user in the process of using his good. The driver does not compensate for the environmental damage caused by using the car.

Cost Estimation

When developing a business plan for a new or existing company, product, or project, planners typically make cost estimates in order to assess whether revenues/benefits will cover costs. This is done in both business and government. Costs are often underestimated, resulting in cost overrun during execution.

Cost-plus pricing, is where the price equals cost plus a percentage of overhead or profit margin.

Manufacturing Costs vs. Non-manufacturing Costs

Manufacturing costs are those costs that are directly involved in manufacturing of products. Examples of manufacturing costs include raw materials costs and charges related to workers. Manufacturing cost is divided into three broad categories:

1. Direct materials cost.
2. Direct labor cost.
3. Manufacturing overhead cost.

Non-manufacturing costs are those costs that are not directly incurred in manufacturing a product. Examples of such costs are salary of sales personnel and advertising expenses. Generally non-manufacturing costs are further classified into two categories:

1. Selling and distribution costs.
2. Administrative costs.

Other Costs

A defensive cost is an environmental expenditure to eliminate or prevent environmental damage. Defensive costs form part of the genuine progress indicator (GPI) calculations.

Labour costs would include travel time, holiday pay, training costs, working clothes, social insurance, taxes on employment &c.

Path cost is a term in networking to define the worthiness of a path.

Direct Materials Cost

Direct materials cost the cost of direct materials which can be easily identified with the unit of production. For example, the cost of glass is a direct materials cost in light bulb manufacturing.

The manufacture of products or goods required material as the prime element. In general, these materials are divided into two categories. These categories are direct materials and indirect materials.

Direct materials are also called productive materials, raw materials, raw stock, stores and only materials without any descriptive title.

Direct Materials Cost Estimation

Steps to estimate the direct material costs:

1. Find the total amount to be produced. This is usually noted as the order size.
2. Calculate the total amount of raw materials required to produce the order size.
3. Multiply that amount by the cost associated with the raw materials.
4. If there is a waste or scrap, its cost should be added to the costs in step 3.
5. If the waste or scrap can be sold at salvage value, this value should be subtracted from the costs in step 4.

Direct Labor Cost

Direct labor cost is a part of wage-bill or payroll that can be specifically and consistently assigned to or associated with the manufacture of a product, a particular work order, or provision of a service. Also, we can say it is the cost of the work done by those workers who actually make the product on the production line.

Determination of the Direct Labor Cost

- Planning the work to be performed.
- Describing the job content of the work, by indicating the skill, knowledge, etc.
- Matching the jobs with the employees.

Usage

The direct labor cost is part of the manufacturing cost.

Calculation of Direct Labor Cost

In the direct labor cost we need to have the job time and wage we will pay it to the worker to calculate the direct labor cost as in this formulation:-

$$\text{Direct labor cost} = \text{job time} \times \text{wage}$$

Wage

The wage is the payment rendered to the worker per hour as a compensation for the work done.

Calculating Job Time

The job time needs to be measured by one of the following ways:

1. time study
2. work sampling
3. Predetermined motion time system

MOH Cost

Manufacturing overhead costs (MOH cost) are all manufacturing costs that are related to the cost object (work in process and then finished goods) but cannot be traced to that cost object in an economically feasible way.

Examples include supplies, indirect materials such as lubricants, indirect manufacturing labor such as plant maintenance and cleaning labor, plant rent, plant insurance, property taxes on the plant, plant depreciation, and the compensation of plant managers.

This cost category is also referred to as Factory overhead cost (FO cost).

Items of the Overhead

Manufacturing overhead includes other costs in manufacturing that are neither direct materials costs nor direct labor costs. It might also be referred as the factory burden or production overhead. Its value is essential for determining the cost of products to be manufactured.

This category of costs includes expenses like:

1. Electricity for the equipment and lighting.
2. Cleaning costs of equipment.
3. Material handling like forklifts
4. Maintenance of the equipment.
5. Wages of employees in the factory who for example work on record keeping and inspection of materials.
6. Computing the operation of the process of manufacturing the product.
7. Insurance
8. Safety and quality cost.

Cost Estimate

A cost estimate is the approximation of the cost of a program, project, or operation. The cost estimate is the product of the cost estimating process. The cost estimate has a single total value and may have identifiable component values. A problem with a cost overrun can be avoided with a credible, reliable, and accurate cost estimate. A cost estimator is the professional who prepares cost estimates. There are different types of cost estimators, whose title may be preceded by a modifier, such as building estimator, or electrical estimator, or chief estimator. Other professional titles may also prepare cost estimates or contribute to cost estimates, such as quantity surveyors, cost engineers, etc. In the US, there were 185,400 cost estimators in 2010. There are around 75,000 professional quantity surveyors working in the UK.

Overview

The U.S. Government Accountability Office (GAO) defines a cost estimate as, "the summation of individual cost elements, using established methods and valid data, to estimate the future costs of a program, based on what is known today." The GAO reports that "realistic cost estimating was imperative when making wise decisions in acquiring new systems." A cost estimate is often needed to support evaluations of project feasibility or funding requirements in support of planning. A cost estimate is often used to establish a budget as the cost constraint for a project or operation.

In project management, project cost management is a major functional division. Cost estimating is one of three activities performed in project cost management.

In cost engineering, cost estimation is a basic activity. A cost engineering reference book has chapters on capital investment cost estimation and operating cost estimation. The fixed capital investment provides the physical facilities. The working capital investment is a revolving fund to keep the facilities operating.

In system, product, or facility acquisition planning, a cost estimate is used to evaluate the required funding and to compare with bids or tenders.

In construction contracting, a cost estimate is usually prepared to submit a bid or tender to compete for a contract award.

In facility maintenance and operation, cost estimates are used to establish funding or budgets.

In an attempt to manage liability risk, some firms avoid the use of the word estimate and instead refer to the estimate as an "Opinion of Probable Cost."

Cost Estimate Types

Various projects and operations have distinct types of cost estimating, which vary in their composition and preparation methods. Some of the major areas include:

- Construction cost (i) detailed construction estimate (ii) abstract construction estimate
- Manufacturing cost
- Software development cost
- Aerospace mission cost
- Resource exploration cost
- Facility operation cost
- Facility maintenance and repair cost
- Facility rehabilitation and renewal cost
- Facility retirement cost

Cost Estimate Classifications

Common cost estimate classifications historically used are

- Order of magnitude
- Detailed estimate
- Preliminary
- Definitive

These correspond to modern published classes 5, 3, and 1, respectively. The U.S. Department of Energy and many others use a system of five classes of estimates:

Estimate class	Name	Purpose	Project definition level
Class 5	Order of magnitude	Screening or feasibility	0% to 2%
Class 4	Intermediate	Concept study or feasibility	1% to 15%
Class 3	Preliminary	Budget, authorization, or control	10% to 40%
Class 2	Substantive	Control or bid/tender	30% to 70%
Class 1	Definitive	Check estimate or bid/tender	50% to 100%

Methods used to prepare the estimates range from stochastic or judgment at early definition to deterministic at later definition. Some estimates use mixed methods.

Cost estimate classifications have been published by ASTM and AACE International. The American Society of Professional Estimators (ASPE) defines estimate levels in the reverse order as Level 1 – Order (Range) of Magnitude, Level 2 – Schematic/Conceptual Design, Level 3- Design Development, Level 4 – Construction Document, and Level 5 – Bid.>." ACostE defines a Class I Estimate as definitive, a Class II Estimate as semi-detailed, and a Class III Estimate as pre-budget.

Other names for estimates of different classes include:

Class 1	Class 3	Class 5
Detailed estimate	Semi-detailed estimate	Conceptual estimate
Final estimate	Scope estimate	Pre-design estimate
Control estimate	Sanction estimate	Preliminary estimate
As-bid estimate		Pre-budget estimate
As-sold estimate		Evaluation estimate
CD estimate	DD estimate	SD estimate
		Parametric estimate
		Rough order-of-magnitude (ROM) estimate
		Very rough order-of-magnitude (VROM) estimate
		SWAG (scientific, wild-ass guess) estimate
		PIDOOMA (pulled-it-directly-out-of-my-ass) estimate

Estimate Quality

Estimate quality refers to the delineation of quality requirements for the estimate. These requirements are set out in accordance with formal quality assurance standards. There may also be other expectations for the estimate which are not specific requirements, but may affect the perceived quality of the estimate. Published quality requirements generally have to do with credibility, accuracy, confidence level, precision, risk, reliability, and validity of the estimate, as well as thoroughness, uniformity, consistency, verification, and documentation.

"The result of bidding without good estimates is certain: jobs that end up with less profit, no profit, or a loss. The bidder ultimately will go out of business; the only question is how long will it take."

Figure 4: Cone of Uncertainty

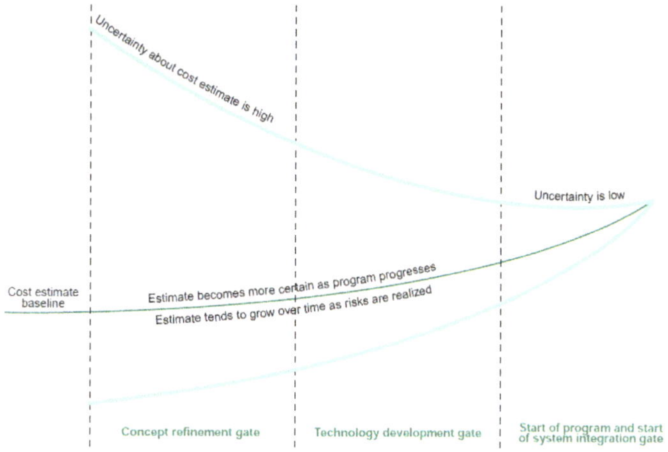

Source: GAO.

Since a cost estimate is the approximation of the cost of a project or operation, then estimate accuracy is a measure of how closely the estimate is able to predict the actual expenditures for the project or operation. This can only be known after the project is completed. If, for example, a project estimate was $1,252,000 for a specific scope and conditions, and at completion the records showed that $1,172,451.26 was expended, the estimate was 6.8% too high. If the project ended up having a different scope or conditions, an unadjusted computation does not fairly assess the estimate accuracy. Predictions of the estimate accuracy may accompany the estimate. Typically this is expressed as a range higher or lower as compared with the point estimate with an expected probability that the actual cost will fall in the range. An example for a definitive estimate might be that the estimate has a -5/+10% range of accuracy with a 90% confidence that the final value will fall in that range. The accuracy of an early estimate relates to the estimate quality. Factors affecting the quality of the estimate include the people who prepared the estimate, how the estimate was prepared, and what was known about the project. For the same project, the range of uncertainty about the total estimate decreases over time, as illustrated in the cone of uncertainty diagram.

High-quality cost estimates can be produced by following a rigor of 12 steps outlined by the U.S. GAO. Detailed documentation is recommended to accompany the estimate. "The documentation addresses the purpose of the estimate, the program background and system description, its schedule, the scope of the estimate (in terms of time and what is and is not included), the ground rules and assumptions, all data sources, estimating methodology and rationale, the results of the risk analysis, and a conclusion about whether the cost estimate is reasonable. Therefore, a good cost estimate—while taking the form of a single number—is supported by detailed documentation that describes how it was derived and how the expected funding will be spent in order to achieve a given objective." This documentation is often titled Basis of Estimate (or BOE). Additional documentation may accompany the estimate, including quantity takeoff documentation and supporting calculations, quotes, etc.

Although the pursuit of cost estimate accuracy should always be encouraged, a study in 2002 found that the estimates used to determine whether important infrastructure should be built were "highly and systematically misleading."

Contingency

A contingency may be included in an estimate to provide for unknown costs which are indicated as likely to occur by experience, but are not identifiable. When using an estimate which has no contingency to set a budget or to set aside funding, a contingency is often added to improve the probability that the budget or funding will be adequate to complete the project. Being unable to complete a project risks public ridicule. The estimate or budget contingency is not intended to compensate for poor estimate quality, and is not intended to fund design growth, owner changes, or anything else unrelated to delivering the scope as defined in the estimate documentation. Generally more contingency is needed for earlier estimates due to the higher uncertainty of estimate accuracy.

Cost Estimating Methods and Best Practices

Estimating methods may vary by type and class of estimate. The method used for most definitive estimates is to fully define and understand the scope, take off or quantify the scope, and apply costing to the scope, which can then be summed to a total cost. Proper documentation and review are also important. Pricing transforms the cost estimate into what the firm wishes to charge for the scope. Early estimates may employ various means of cost modeling. The basic characteristics of effective estimating include: clear identification of task, broad participation in preparing estimates, availability of valid data, standardized structure for the estimate, provision for program uncertainties, recognition of inflation, recognition of excluded costs, independent review of estimates, and revision of estimates for significant program changes. Application of best practices helps ensure a high-quality estimate. "Certain best practices should be followed if accurate and credible cost estimates are to be developed. These best practices represent an overall process of established, repeatable methods that result in high-quality cost estimates that are comprehensive and accurate and that can be easily and clearly traced, replicated, and updated."

Tools that may be part of costs estimation are cost indexes. These factors promote time adjustment of capital costs, following changes in technology, availability of materials and labor, and inflation. Due to the inherent unavailability of up-to-date cost literature, several inflation or cost indexes are available.

Construction Cost Estimates

Estimates for the cost of facility construction are a major part of the cost estimate domain. A construction general contractor or subcontractor must normally prepare definitive cost estimates to prepare bids in the construction bidding process to compete for award of the contract. Although many estimators participate in the bidding and procurement processes, those are not a necessary function of cost estimate preparation. Earlier estimates are prepared by differing methods by estimators and others to support the planning process and to compare with bids. One way to make those estimates is by determining the resources needed (e.g., the amount of construction material quantities that are required) and then multiplying the estimated construction material quantities by the corresponding unit cost. One advantage of making estimates in this way is that it allows for the segregation of quantities and costs. This way they can be updated separately as new information becomes available. They can also be tracked separately allowing decision makers to make better decisions about the project during its conceptual phase.

Definitive Estimates (Class 1)

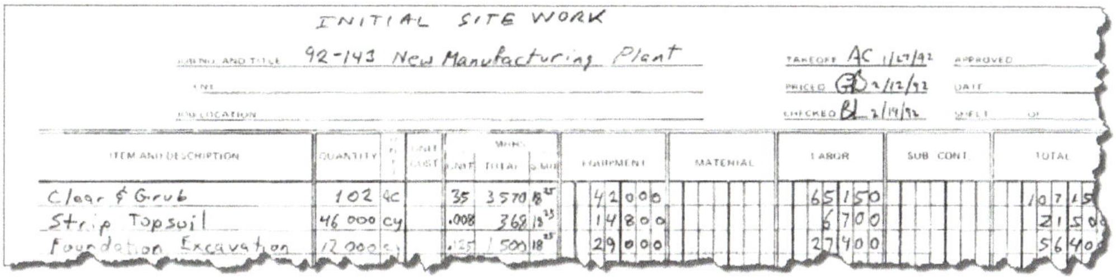

A handwritten spreadsheet.

A definitive estimate is prepared from fully designed plans and specifications (or nearly so), preferably what are called contract documents (CD). The contract documents also establish the Scope of Work (SOW). The standard method is to review and understand the design package and take off (or perform a quantity survey of) the project scope by itemizing it into line items with measured quantities. RSMeans refers to this as, "Scope out the project," and, "Quantify." Some jurisdictions or areas of practice define the itemization and measuration in certain terms, such as RICS and may have specific rules for development of a Bill of quantities, or BOQ. The ASPE proposes a best practice standard method for the quantity survey. This includes using the Construction Specifications Institute Uniform Numbering System (MasterFormat) to ensure that all work is accounted for.

Then costs are applied to the quantified line items. This may be called costing or pricing. In estimating for contracting, the cost is what something costs you to build, and price is what you charge another party for building it. RSMeans refers to this as, "Price the quantities." ASPE recommends the "quantity times material and labor costs format" for the compilation of the estimate. This format is illustrated in the handwritten spreadsheet sample. For labor, the estimator should, "Determine basic production rates and multiply them by the units of work to determine total hours for the work." and then multiply the hours by the per hour average labor cost. Labor burdens, material costs, construction equipment costs, and, if applicable, subcontractor costs are also extended on the estimate detail form. Other costs and pricing are added, such as overhead, profit, sales or use taxes, payment and performance bonds, escalation, and contingency.

The costs which are applied to the line-item quantities may come from a cost book (either internal or external) or cost database. For construction contractors or construction managers it is important to track and compile past data of trends, completed projects, production factors, equipment changes, and various labor markets.

Description	Size	Unit	Man-hours Norm	Productivity Factor	Project Man-hours Rate
Rigid Steel Conduit	1/2"	lf	0.045	1.200	0.054
Rigid Steel Conduit	3/4"	lf	0.050	1.200	0.060
Rigid Steel Conduit	1"	lf	0.055	1.200	0.066
Rigid Steel Conduit	1-1/2"	lf	0.075	1.200	0.090
Rigid Steel Conduit	2"	lf	0.085	1.200	0.102

For normal installation conditions and projects, man-hours include:
receiving, unloading, unpacking, and moving to installation location
drawing review, layout, and marking locations
measure, cut, and thread
temporary supporting, climbing, etc.
standby time, cleanup time, etc.
Excludes:
box connections and fittings, but includes couplings
permanent hangers, supports or fastenings

Sample labor norms

The labor requirements are often the most variable and are a primary focus of construction cost estimators. The labor hours required to construct each installation item are calculated by using a man-hour rate times the take-off quantity (a similar method is to divide the take-off quantity by the production rate). Many estimators use a man-hour norm reference for standard man-hours and apply an adjustment factor for project or task conditions, location, methods, equipment, labor skill, etc. to adjust for the anticipated effect on labor.

Direct costs are itemized for all necessary parts of the project. Direct costs are all of the costs which can be attributed directly to the project. Direct costs include costs for general requirements (Division 1 of MasterFormat), which includes such items as project management and coordination, quality control, temporary facilities and controls, cleaning and waste management. Direct costs may also include the costs of project planning, investigation, studies, and design; land or right of way acquisition, and other non-construction costs. Usually, a subtotal of total direct costs is provided in the estimate.

Provisions are made for Indirect costs in addition to the direct costs. Indirect costs include overhead, profit, sales or use taxes, payment and performance bonds, escalation, and contingency. Profit is cost to the buyer, but is not a cost to the provider, rather a projection of anticipated income.

A well documented cost estimate includes a Basis of Estimate (BOE), which describes the scope basis, pricing basis, methods, assumptions, inclusions, and exclusions.

Order-of-Magnitude Estimates (Class 5)

An order-of-magnitude estimate is prepared when little or no design information is available for the project. It is called order of magnitude because that may be all that can be determined at an early stage. In other words, perhaps we can only determine that it is of a 10,000,000 magnitude as opposed to a 1,000,000 magnitude. Various techniques are employed for these estimates, including experience and judgment, historical values and charts, rules of thumb, and simple mathematical calculations. Factor estimating is one of the more popular methods. This involves taking the known cost of a similar facility and factoring the cost for size, place, and time. Cost modeling is another common technique. In cost modeling the estimator models the various parameters of the facility and applies costs to the derived scope.

Building estimators or architects may use the Uniformat system of breaking down the building into functional systems or assemblies during the schematic design (SD) phase of planning and design. The RSMeans Square Foot Costs book organizes building costs according to the 7 divisions of the UNIFORMAT II classification system. The 7 divisions are:

- A Substructure
- B Shell
- C Interiors
- D Services
- E Equipment & Furnishings
- F Special Construction
- G Building Site Work

Construction Estimating Software

Construction cost estimating software is computer software designed for contractors to estimate construction costs for a specific project. An cost estimator will typically use estimating software to estimate their bid price for a project, which will ultimately become part of a resulting construction contract. Some architects, engineers, construction managers, and others may also use cost estimating software to prepare cost estimates for purposes other than bidding.

History

Traditional Methods

Construction contractors usually prepare bids or tenders to compete for a contract award for a project. To prepare the bid, first a cost estimate is prepared to determine the costs and then establish the price(s). This involves reviewing the project's plans and specifications to produce a take off or quantity survey, which is a listing of all the materials and items of work required for a construction project by the construction documents. Together with prices for these components, the measured quantities are the basis for calculation of the direct cost. Indirect costs and profit are added to arrive at a total amount.

Spreadsheets

Cost estimators used columnar sheets of paper to organize the take off and the estimate itself into rows of items and columns containing the description, quantity and the pricing components. Some of these were similar to accounting ledger paper. They became known as green sheets or spreadsheets.

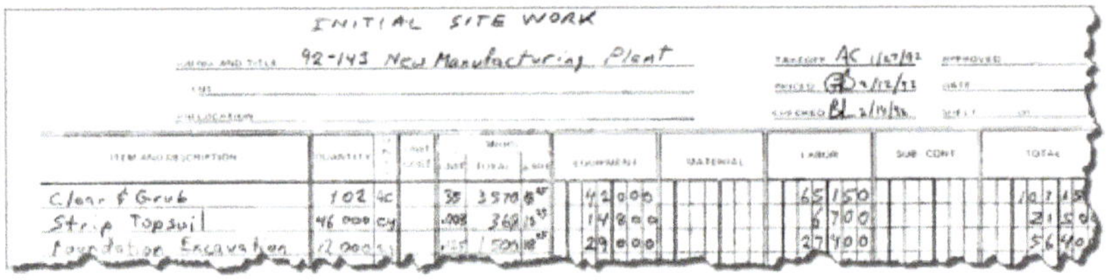

A handwritten spreadsheet.

Work Item	Vendor	Labor	Equipment	Materials	Subcontr.	Subtotal	Markup %	Markup	Total
Permits/Fees	City of Los Angeles				$1,500.00	$1,500.00		$0.00	$1,500.00
Excavation		$6,000.00	$8,000.00	$500.00		$14,500.00	15.00%	$2,175.00	$16,675.00
Utilities		$3,500.00	$2,500.00	$2,750.00	$1,000.00	$9,750.00	15.00%	$1,462.50	$11,212.50
Water Well						$0.00		$0.00	$0.00
Septic Tank						$0.00		$0.00	$0.00
Foundation	Connie's Concrete				$3,500.00	$3,500.00	5.00%	$175.00	$3,675.00
Concrete Flatwork	Connie's Concrete				$1,900.00	$1,900.00	5.00%	$95.00	$1,995.00
Framing		$3,500.00	$1,500.00	$9,000.00		$14,000.00	15.00%	$2,100.00	$16,100.00
Roofing	Robert's Roofing				$3,500.00	$3,500.00	5.00%	$175.00	$3,675.00
Windows/Ext.Doors	Wally's Windows				$8,000.00	$8,000.00	5.00%	$400.00	$8,400.00
Garage Door	Gary's Garage Doors				$2,250.00	$2,250.00	5.00%	$112.50	$2,362.50
Siding						$0.00		$0.00	$0.00
Electrical	Ernie's Electric				$18,500.00	$18,500.00	5.00%	$925.00	$19,425.00
Plumbing	Mac's Mechanical				$16,500.00	$16,500.00	5.00%	$825.00	$17,325.00
HVAC	Mac's Mechanical				$23,000.00	$23,000.00	5.00%	$1,150.00	$24,150.00
Insulation		$3,500.00		$1,000.00		$4,500.00		$0.00	$4,500.00
Masonry	Mason's Masonry				$14,500.00	$14,500.00	5.00%	$725.00	$15,225.00
Drywall	Doug's Drywall				$12,500.00	$12,500.00	5.00%	$625.00	$13,125.00
Interior Trim	Doug's Drywall				$9,000.00	$9,000.00	5.00%	$450.00	$9,450.00
Painting	Paul's Painting				$13,500.00	$13,500.00	5.00%	$675.00	$14,175.00
Floor Coverings	Carl's Carpets				$16,500.00	$16,500.00	5.00%	$825.00	$17,325.00
Cabinets	Ken's Kabinets				$22,500.00	$22,500.00	5.00%	$1,125.00	$23,625.00
Appliances	Abby's Appliances	$2,500.00		$11,500.00		$14,000.00	15.00%	$2,100.00	$16,100.00
Landscaping	Sonny's Sodding				$2,750.00	$2,750.00	5.00%	$137.50	$2,887.50
Overhead Costs		$10,000.00				$10,000.00	20.00%	$2,000.00	$12,000.00
Other						$0.00		$0.00	$0.00
						$0.00		$0.00	$0.00
						$0.00		$0.00	$0.00
						$0.00		$0.00	$0.00
						$0.00		$0.00	$0.00
						$0.00		$0.00	$0.00
TOTALS						$236,650.00	7.71%	$18,257.50	$254,907.50

A basic estimating spreadsheet.

With the advent of computers in business, estimators began using spreadsheet applications like VisiCalc, Lotus 1-2-3, and Microsoft Excel to duplicate the traditional tabular format, while automating redundant mathematical formulas.

Many construction cost estimators continue to rely primarily upon manual methods, hard copy documents, and/or electronic spreadsheets such as Microsoft Excel. While spreadsheets are relatively easy to master and provide a means to create and report a construction cost estimate and or cost models, their benefit comes largely from their ability to partially relieve estimators of mundane calculations. Accuracy, however, is not necessarily improved and productivity is not maximized. For example, data entry remains tedious and prone to error, formula errors are common, and collaboration and information sharing are limited.

Commercial Estimating Software

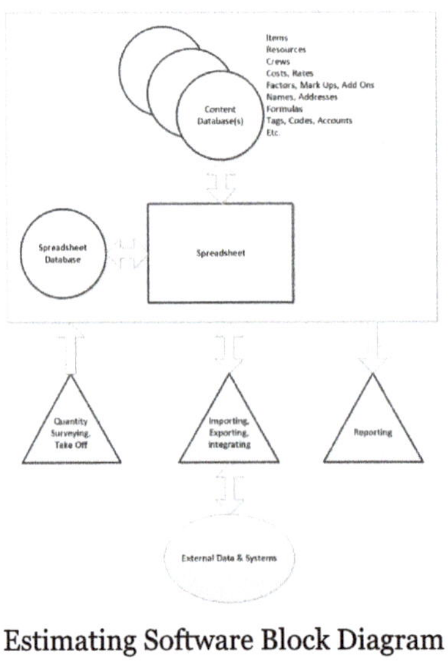

Estimating Software Block Diagram

As cost estimators came to rely heavily on spreadsheets, and the formulas within the spreadsheets became more complex, spreadsheet errors became more frequent. These were typically formula errors and cell-reference errors which would often lead to cost overruns. As a result, commercial cost estimating software applications were originally created to overcome these errors by using hard-coded formulas and data structures. Other benefits include the use of reference databases for costs and other data, predictable and professional looking reports, speed, accuracy, and overall process standardization.

As cost estimating programs became more and more popular over the years, more advanced features, such as saving data for reuse and trade-specific calculations, have become available. For example, programs that are designed for building construction, include libraries and program features for traditional builders. In sharp contrast, programs that are designed for civil construction, include libraries and program features for roadway, utility, and bridge builders.

Sophisticated, cost estimating and Efficient Project Delivery Software systems are also available to integrate various construction delivery methods such as Integrated Project Delivery, Job Order Contracting, and others (IDIQ, JOC, SABER...) simultaneously and securely. These systems enable cost estimators and project managers to collaboratively work with multiple projects, multiple estimates, and multiple contracts. A 'short list' of additional capabilities includes the ability to work with multiple cost books/guides/UPBs, track project status, automatically compare estimates, easily copy/paste, clone, and reuse estimates, and integrated sophisticated visual estimating and quantity take-off (QTO) tools. Owners, contractors, architects and engineers are moving to advanced cost estimating and management systems, and many oversight groups such are beginning to also require their use. The level of collaboration, transparency, and information re-use enabled by Cost Estimating and Efficient Project Delivery Software drives 15-25%+ reductions in procurement cycles, six to ten times faster estimating, reduce overall project times, as a significant reduction in change orders and the virtual elimination of contract related legal disputes.

Typical Features

Three functions prove to be the most critical when buying cost estimating software:

- Takeoff software - this provides for measurement from paper or electronic plans
- Built-in cost databases - this provides reference cost data which may be your own or may come from a commercial source, such as RS Means
- Estimating worksheets - these are the spreadsheets where the real work takes place, supported by calculations and other features

Other typical features include:

- Item or Activity List: All estimating software applications will include a main project window that outlines the various items or activities that will be required to complete the specified project. More advanced programs are capable of breaking an item up into sub-tasks, or sub-levels. An outline view of all of the top-level and sub-level items provides a quick and easy way to view and navigate through the project.
- Resource Costs: Resources consist of labor, equipment, materials, subcontractors, truck-

ing, and any other cost detail items. Labor and equipment costs are internal crew costs, whereas all other resource costs are received from vendors, such as material suppliers, subcontractors, and trucking companies. Labor costs are usually calculated from wages, benefits, burden, and workers compensation. Equipment costs are calculated from purchase price, taxes, fuel consumption, and other operating expenses.

- Item or Activity Detail: The detail to each item includes all of the resources required to complete each activity, as well as their associated costs. Production rates will automatically determine required crew costs.

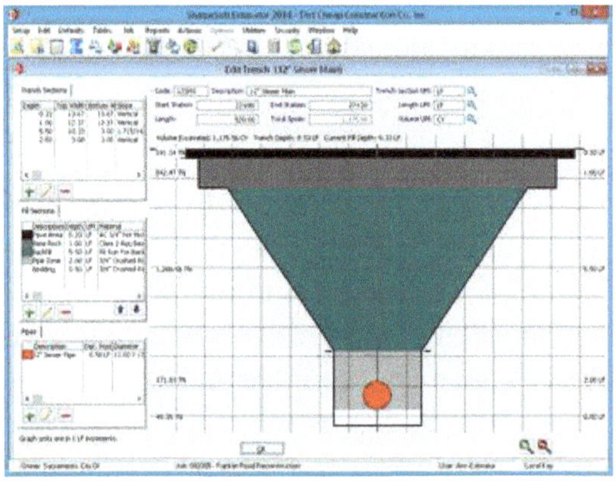

The SharpeSoft Estimator, displaying an example of a trench calculation.

- Calculations: Most estimating programs have built-in calculations ranging from simple length, area, and volume calculations to complex industry-specific calculations, such as electrical calculations, utility trench calculations, and earthwork cut and fill calculations.
- Markups: Every program will allow for cost mark-ups ranging from flat overall mark-ups to resource-specific mark-ups, mark-ups for general administrative costs, and bonding costs.
- Detailed Overhead: Indirect costs, such as permits, fees, and any other overall project costs can be spread to billable project items.
- Closeout Window: Many estimating programs include a screen for manually adjusting bid prices from their calculated values.
- Reporting: Project reports typically include proposals, detail reports, cost breakdown reports, and various charts and graphs.
- Exporting: Most software programs can export project data to other applications, such as spreadsheets, accounting software, and project management software.
- Job History: Storing past projects is a standard feature in most estimating programs.

Cost Engineering

Cost engineering is "the engineering practice devoted to the management of project cost, involving such activities as estimating, cost control, cost forecasting, investment appraisal and risk analysis." "Cost Engineers budget, plan and monitor investment projects. They seek the optimum balance between cost, quality and time requirements."

Skills and knowledge of cost engineers are similar to those of quantity surveyors. In many industries, cost engineering is synonymous with project controls. As the title "engineer" has legal requirements in many jurisdictions (i.e. Texas, Canada), the cost engineering discipline is often renamed to project controls.

A cost engineer is "an engineer whose judgment and experience are utilized in the application of scientific principles and techniques to problems of estimation; cost control; business planning and management science; profitability analysis; project management; and planning and scheduling."

Overview

One key objective of cost engineering is to arrive at accurate cost estimates and schedules and to avoid cost overruns and schedule slips. Cost engineering goes beyond preparing cost estimates and schedules by helping manage resources and supporting assessment and decision making. "The discipline of 'cost engineering' can be considered to encompass a wide range of cost-related aspects of engineering and programme management, but in particular cost estimating,cost analysis/cost assessment, design-to-cost, schedule analysis/planning and risk assessment." The broad array of cost engineering topics represent the intersection of the fields of project management, business management, and engineering. Most people have a limited view of what engineering encompasses. The most obvious perception is that engineering addresses technical issues such as the physical design of a structure or system. However, beyond the physical manifestation of a design of a structure or system (for example, a building), there are other dimensions to consider such as the money, time, and other resources that were invested in the creation of the building. Cost engineers refer to these investments collectively as "costs".

Cost engineering, then, can be considered an adjunct of traditional engineering. It recognizes and focuses on the relationships between the physical and cost dimensions of whatever is being "engineered". Cost engineering is most often taught at universities as part of construction engineering, engineering management, civil engineering, and related curricula because it is most often practiced on engineering and construction capital projects. Engineering economics is a core skill and knowledge area of cost engineering.

AACE International "is dedicated to the tenets of furthering the concepts of Total Cost Management and Cost Engineering. Total Cost Management is the effective application of professional and technical expertise to plan and control resources, costs, profitability and risk. Simply stated, it is a systematic approach to managing cost throughout the life cycle of any enterprise, program, facility, project, product or service. This is accomplished through the application of cost engineering and cost management principles, proven methodologies and the latest technology in support of the management process. Total Cost Management is that area of engineering practice where engineering judgment and experience are utilized in the application of scientific principles and techniques to problems of business and program planning; cost estimating; economic and financial analysis; cost engineering; program and project management; planning and scheduling; and cost and schedule performance measurement and change control. In summary, the list of practice areas ... are collectively called cost engineering; while the "process" through which these practices are applied is called total cost management or TCM.

History

Cost engineering is a field of engineering practice that began in the 1950s (AACE International was founded in 1956). The skills and knowledge areas of cost engineers are similar to those of quantity surveyors. AACE International is one of many international engineering organizations representing practitioners in these fields. The International Cost Engineering Congress (ICEC) was founded in 1976 as a Worldwide Confederation of Cost Engineering, Quantity Surveying and Project Management Societies.

In 2006, AACE published the Total Cost Management (TCM) Framework which outlines an integrated process for applying the skills and knowledge of cost engineering. This has also been called the world's first process for portfolio, program and project management.

Professional Titles or Positions in Cost Engineering

"Cost engineering practitioners tend to be: a) specialized in function (e.g., cost estimating, planning and scheduling, etc.); b) focused on either the asset management or project control side of the TCM process; and c) focused on a particular industry (e.g., engineering and construction, manufacturing, information technology, etc) or asset type (e.g., chemical process, buildings, software, etc.)... They may work for the business that owns and operates the asset (emphasis on economics and analysis), or they may work for the contractor that executes the projects (emphasis on planning and control)."

Some titles or positions in the Cost Engineering practice include:

- Claims and Changes Specialist
- Construction Estimator
- Construction Manager
- Contract Management Specialist
- Cost Analyst
- Cost Engineer
- Cost Estimator (or Estimator)
- Planner/Scheduler (or Scheduling Engineer)
- Pre-Construction Manager
- Project Controls Engineer
- Project Controls Manager
- Project Controls
- Project Controller
- Project Manager

Rate Analysis

At various stages in the project management, we need to know

- how much is cost of executing unit amount of the work
- how many equipment or labour are required to execute unit amount of an item of work.

These things are required for the purpose of eliminating cost of construction, material planning, labour and equipment allocation in the scheduling etc. Obtaining the cost of unit amount of an item is called rate analysis. To obtain the rate of an item, generally following cost are considered.

- cost of material
- cost of labour
- cost of equipments, plant tools etc.
- overhead cost
- profit

The costs of these components are determined from the area survey, schedule of rates published by central and state government; transportation charges based on the location of the area. Also we need to know how much is required for execution of the work. Similarly number of labour and equipment. These things are bared on both the heuristics and engineering design. Not only specification is important for obtaining the cost of the work, but one has also to include quality mentioned during construction we shall discuss this pointing detail through case study.

a. Cost of material – To obtain the cost of material in the unit amount of an item, first specification is studied carefully. Quantity of material required is calculated and multiplied by the cost to obtain the unit cost. Cost includes freight, transportation, sales tax, insurance, as well as profit. An estimator has to do his market survey and the area survey before calculating the unit cost of the material.

b. Cost of labour – Productivity of the labour is very important aspect in calculation of labour cost. Productivity of the labour depends upon the type of work, training imparted to the labourer, nature of work, time schedule of the work, etc. For example, a mason can do more plastering standing on the floor than working at a height on the scaffolding. Initially, it takes time to learn the work and gradually productivity increases, which is shown by the learning curve as follows –

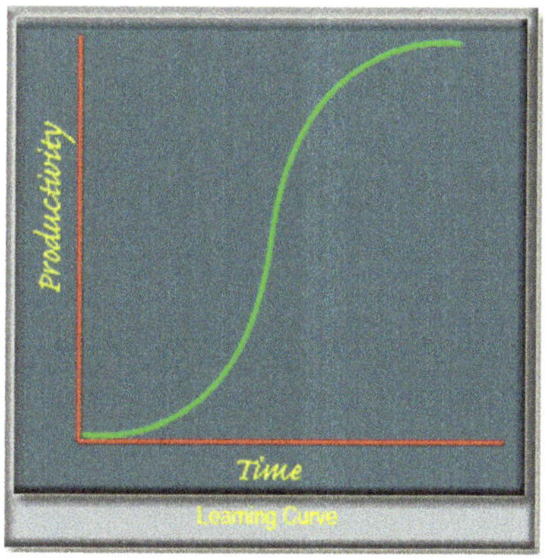

For specific works, different types of labourers, such as skilled and unskilled labour, mason, carpenter, mazdoor, etc are required to execute the work. Unit amount of work is obtained based on experience; and their cost is obtained from the local survey or government wage rules.

c. Cost of equipment - Calculation of work-hour used for equipment is dependent upon the type of equipment utility. Some equipments are used for specific purpose where as some of it are used for different work simultaneously. Generally equipment used individually is calculated for each item of work. For these equipments, it is not possible to calculate the cost; the equipment charge can be added as overhead. For example, installation of a batching plant requires special effort and it is not easy to distribute its initial cost on the concreting. We can incorporate such kind of work as establishment charges or overhead.

d. Overhead cost – During the period of construction, a company has to maintain an office or depreciation of its equipments may take place. This is usually taken as 2-5% of unit rate.

Part of the overhead is directly related to the construction of an item of work. Such expenses include –

i. Repair and depreciation of tool and plants

ii. Salaries of supervisors, etc; who are indirectly involved in the work

iii. Temporary godown and sheds to store the material at the site

iv. Lighting arrangements

v. Labour welfare

vi. Safety scheme

vii. Small tools used for work

Such expenses are incurred only when work is under execution. If there is no work, no expense is required.

Other parts of overhead are permanent in nature. These expenses are throughout the year. These expenses include –

i. Salaries of office staff,

ii. Printing and stationary charges,

iii. Electricity and rental charges,

iv. Communication expenses such as telephone, e-mail, etc,

v. Traveling, etc.

In bigger companies, these expenses are higher compared to smaller contractors.

Water surcharged: -1% water surcharge is generally added in the rate analysis.

e. Profit – The profit targeted in construction works is between 7-15%. In large jobs, profit is low where as profit considered is higher in smaller jobs. We shall illustrate the calculation of rate analysis through the examples –

i. Obtain the unit rate of production of M20 concrete.

ii. Obtain the cost of brick masonry for one cubic meter, for one brick thick wall.

Prime Cost

The owner or the department, in the case of government construction, supplies many Items during construction. The contractor's job is to install such items. Examples of such items are door and windows, water supply and sanitary fittings, fan and other electrical appliances or Internet cables, etc. Cost of these items at the shop is called Prime Cost. Let an organization provide the fan to the contractor. The owner buys the fan at the cost of Rs. 700. Thus the prime cost of the fan is Rs. 700/-. The prime cost of an item is obtained based on the cash memo (Purchase bill) of the owner.

In many cases, it is possible to give exact specification of certain items during the preparation of estimate or contract awarding. The owner can specify such items and the contractor purchases the item. The payment is made based on the prime cost. The government departments in our country follow such practices. CPWD specifies various rules regarding handling of such an item.

References

- Flyvbjerg, Bent (2002). "Underestimating Costs in Public Works Projects: Error or Lie?". Journal of American Planning Association. 68 (3): 279–295
- Guide to the Project Management Body of Knowledge (PMBOK Guide) Third Edition, An American National Standard, ANSI/PMI 99-001-2004, Project Management Institute, Inc, 2004, ISBN 1-930699-45-X
- Garold D. Oberlender and Steven M. Trost, "Predicting Accuracy of Early Cost Estimates Based on Estimate Quality", Journal of Construction Engineering and Management / Volume 127/ Issue 3/ TECHNICAL PAPERS, American Society of Civil Engineers, Abstract
- O'Sullivan, Arthur; Sheffrin, Steven M. (2003). Economics: Principles in Action. Upper Saddle River, New Jersey 07458: Pearson Prentice Hall. p. 16. ISBN 0-13-063085-3
- García de Soto, B.; Adey, B. T.; Fernando, D. (2014). "A process for the development and evaluation of preliminary construction material quantity estimation models using backward-elimination-regression and neural networks". Journal of Cost Analysis and Parametrics. 7 (3): 180–218. doi:10.1080/1941658X.2014.984880

4

Project Planning and Management

Project management is essential to produce or develop a service or a product. Since a project is a temporary undertaking, normally time-bound, its management assumes necessity to achieve the desired goal. Project planning, a part of project management, uses schedules assess progress. This chapter has been carefully written to provide an easy understanding of the varied facets of project planning and management.

Project Management

Project management is the discipline of initiating, planning, executing, controlling, and closing the work of a team to achieve specific goals and meet specific success criteria. A project is a temporary endeavor designed to produce a unique product, service or result with a defined beginning and end (usually time-constrained, and often constrained by funding or deliverables) undertaken to meet unique goals and objectives, typically to bring about beneficial change or added value. The temporary nature of projects stands in contrast with business as usual (or operations), which are repetitive, permanent, or semi-permanent functional activities to produce products or services. In practice, the management of these two systems is often quite different, and as such requires the development of distinct technical skills and management strategies.

The primary challenge of project management is to achieve all of the project goals within the given constraints. This information is usually described in a user or project manual, which is created at the beginning of the development process. The primary constraints are scope, time, quality and budget. The secondary — and more ambitious — challenge is to optimize the allocation of necessary inputs and integrate them to meet pre-defined objectives.

History

Roman soldiers building a fortress, Trajan's Column 113 AD

Until 1900, civil engineering projects were generally managed by creative architects, engineers, and master builders themselves, for example, Vitruvius (first century BC), Christopher Wren (1632–1723), Thomas Telford (1757–1834) and Isambard Kingdom Brunel (1806–1859). In the 1950s organizations started to systematically apply project-management tools and techniques to complex engineering projects.

Henry Gantt (1861–1919), the father of planning and control techniques

As a discipline, project management developed from several fields of application including civil construction, engineering, and heavy defense activity. Two forefathers of project management are Henry Gantt, called the father of planning and control techniques, who is famous for his use of the Gantt chart as a project management tool (alternatively *Harmonogram* first proposed by Karol Adamiecki); and Henri Fayol for his creation of the five management functions that form the foundation of the body of knowledge associated with project and program management. Both Gantt and Fayol were students of Frederick Winslow Taylor's theories of scientific management. His work is the forerunner to modern project management tools including work breakdown structure (WBS) and resource allocation.

The 1950s marked the beginning of the modern project management era where core engineering fields come together to work as one. Project management became recognized as a distinct discipline arising from the management discipline with engineering model. In the United States, prior to the 1950s, projects were managed on an ad-hoc basis, using mostly Gantt charts and informal techniques and tools. At that time, two mathematical project-scheduling models were developed. The "Critical Path Method" (CPM) was developed as a joint venture between DuPont Corporation and Remington Rand Corporation for managing plant maintenance projects. And the "Program Evaluation and Review Technique" or PERT, was developed by the United States Navy in conjunction with the Lockheed Corporation and Booz Allen Hamilton as part of the Polaris missile submarine program.

PERT and CPM are very similar in their approach but still present some differences. CPM is used for projects that assume deterministic activity times; the times at which each activity will be carried out are known. PERT, on the other hand, allows for stochastic activity times; the times at which each activity will be carried out are uncertain or varied. Because of this core difference, CPM and PERT are used in different contexts. These mathematical techniques quickly spread into many private enterprises.

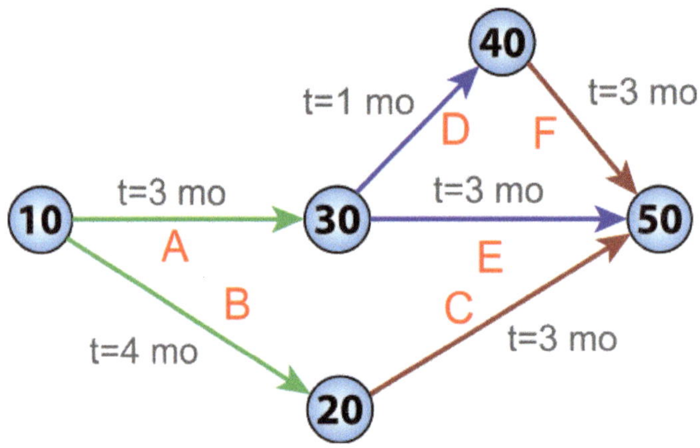

PERT network chart for a seven-month project with five milestones

At the same time, as project-scheduling models were being developed, technology for project cost estimating, cost management and engineering economics was evolving, with pioneering work by Hans Lang and others. In 1956, the American Association of Cost Engineers (now AACE International; the Association for the Advancement of Cost Engineering) was formed by early practitioners of project management and the associated specialties of planning and scheduling, cost estimating, and cost/schedule control (project control). AACE continued its pioneering work and in 2006 released the first integrated process for portfolio, program and project management (Total Cost Management Framework).

The International Project Management Association (IPMA) was founded in Europe in 1967, as a federation of several national project management associations. IPMA maintains its federal structure today and now includes member associations on every continent except Antarctica. IPMA offers a Four Level Certification program based on the IPMA Competence Baseline (ICB). The ICB covers technical, contextual, and behavioral competencies.

In 1969, the Project Management Institute (PMI) was formed in the USA. PMI publishes *A Guide to the Project Management Body of Knowledge* (PMBOK Guide), which describes project management practices that are common to "most projects, most of the time." PMI also offers multiple certifications.

Approaches

There are a number of approaches for managing project activities including lean, iterative, incremental, and phased approaches.

Regardless of the methodology employed, careful consideration must be given to the overall project objectives, timeline, and cost, as well as the roles and responsibilities of all participants and stakeholders.

The Traditional Approach

A traditional phased approach identifies a sequence of steps to be completed. In the "traditional approach", five developmental components of a project can be distinguished (four stages plus control):

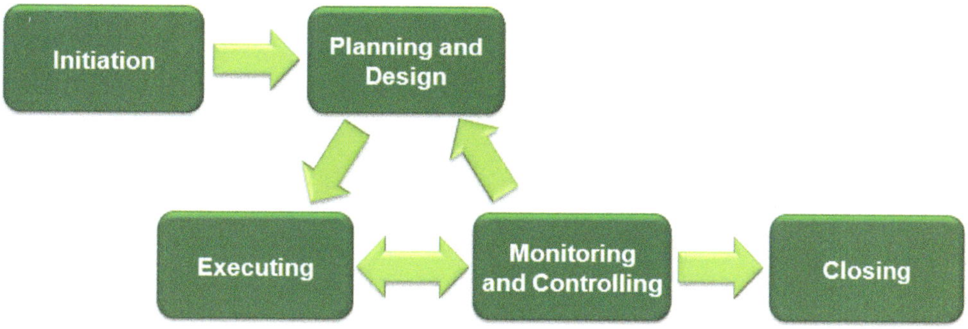

Typical development phases of an engineering project

1. initiation
2. planning and design
3. execution and construction
4. monitoring and controlling systems
5. completion or closing

Many industries use variations of these project stages and it is not uncommon for the stages to be renamed in order to better suit the organization. For example, when working on a brick-and-mortar design and construction, projects will typically progress through stages like pre-planning, conceptual design, schematic design, design development, construction drawings (or contract documents), and construction administration. In software development, this approach is often known as the waterfall model, i.e., one series of tasks after another in linear sequence. In software development many organizations have adapted the Rational Unified Process (RUP) to fit this methodology, although RUP does not require or explicitly recommend this practice. Waterfall development works well for small, well-defined projects, but often fails in larger projects of undefined and ambiguous nature. The Cone of Uncertainty explains some of this as the planning made on the initial phase of the project suffers from a high degree of uncertainty. This becomes especially true as software development is often the realization of a new or novel product. In projects where requirements have not been finalized and can change, requirements management is used to develop an accurate and complete definition of the behavior of software that can serve as the basis for software development. While the terms may differ from industry to industry, the actual stages typically follow common steps to problem solving—"defining the problem, weighing options, choosing a path, implementation and evaluation."

PRINCE2

PRINCE2 is a structured approach to project management released in 1996 as a generic project management method. It combines the original PROMPT methodology (which evolved into the PRINCE methodology) with IBM's MITP (managing the implementation of the total project) methodology. PRINCE2 provides a method for managing projects within a clearly defined framework.

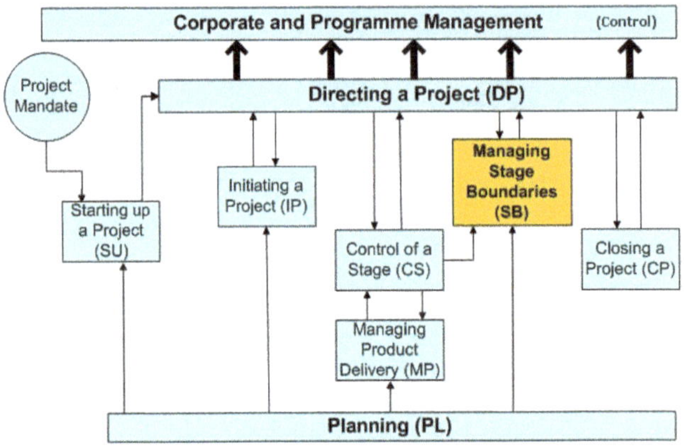

The PRINCE2 process model

PRINCE2 focuses on the definition and delivery of products, in particular their quality requirements. As such, it defines a successful project as being output-oriented (not activity- or task-oriented) through creating an agreed set of products that define the scope of the project and provides the basis for planning and control, that is, how then to coordinate people and activities, how to design and supervise product delivery, and what to do if products and therefore the scope of the project has to be adjusted if it does not develop as planned.

In the method, each process is specified with its key inputs and outputs and with specific goals and activities to be carried out to deliver a project's outcomes as defined by its Business Case. This allows for continuous assessment and adjustment when deviation from the Business Case is required.

PRINCE2 provides a common language for all participants in the project. The governance framework of PRINCE2 – its roles and responsibilities – are fully described and require tailoring to suit the complexity of the project and skills of the organization.

Critical Chain Project Management

Critical chain project management (CCPM) is a method of planning and managing project execution designed to deal with uncertainties inherent in managing projects, while taking into consideration limited availability of resources (physical, human skills, as well as management & support capacity) needed to execute projects.

CCPM is an application of the theory of constraints (TOC) to projects. The goal is to increase the flow of projects in an organization (throughput). Applying the first three of the five focusing steps of TOC, the system constraint for all projects is identified as are the resources. To exploit the constraint, tasks on the critical chain are given priority over all other activities. Finally, projects are planned and managed to ensure that the resources are ready when the critical chain tasks must start, subordinating all other resources to the critical chain.

The project plan should typically undergo resource leveling, and the longest sequence of resource-constrained tasks should be identified as the critical chain. In some cases, such as managing contracted sub-projects, it is advisable to use a simplified approach without resource leveling.

In multi-project environments, resource leveling should be performed across projects. However, it is often enough to identify (or simply select) a single "drum". The drum can be a resource that acts as a constraint across projects, which are staggered based on the availability of that single resource.

One can also use a "virtual drum" by selecting a task or group of tasks (typically integration points) and limiting the number of projects in execution at that stage.

Process-based Management

The incorporation of process-based management has been driven by the use of Maturity models such as the OPM3 and the CMMI (capability maturity model integration; see this example of a predecessor) and ISO/IEC15504 (SPICE – software process improvement and capability estimation). Unlike SEI's CMM, the OPM3 maturity model describes how to make project management processes capable of performing successfully, consistently, and predictably in order to enact the strategies of an organization.

Lean Project Management

Lean project management uses the principles from lean manufacturing to focus on delivering value with less waste and reduced time.

Project Production Management (PPM)

Project Production Management uses principles from operations research, industrial engineering and queuing theory to organize work activities in major capital projects to optimize project delivery.

Extreme Project Management/Megaproject

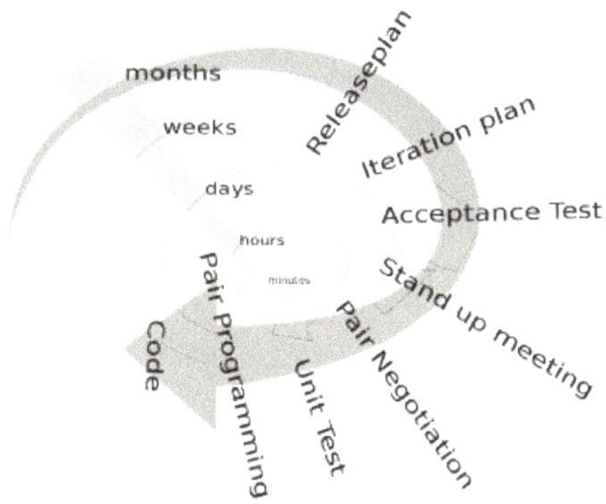

Planning and feedback loops in Extreme programming (XP) with the time frames of the multiple loops.

In critical studies of project management it has been noted that several PERT based models are not well suited for the multi-project company environment of today. Most of them are aimed at very large-scale, one-time, non-routine projects, and currently all kinds of management are expressed in terms of projects.

Using complex models for "projects" (or rather "tasks") spanning a few weeks has been proven to cause unnecessary costs and low maneuverability in several cases. The generalization of Extreme

Programming to other kinds of projects is extreme project management, which may be used in combination with the process modeling and management principles of human interaction management.

Benefits Realization Management

Benefits realization management (BRM) enhances normal project management techniques through a focus on outcomes (the benefits) of a project rather than products or outputs, and then measuring the degree to which that is happening to keep a project on track. This can help to reduce the risk of a completed project being a failure by delivering agreed upon requirements/outputs but failing to deliver the benefits of those requirements.

In addition, BRM practices aim to ensure the alignment between project outcomes and business strategies. The effectiveness of these practices is supported by recent research evidencing BRM practices influencing project success from a strategic perspective across different countries and industries.

An example of delivering a project to requirements might be agreeing to deliver a computer system that will process staff data and manage payroll, holiday and staff personnel records. Under BRM the agreement might be to achieve a specified reduction in staff hours required to process and maintain staff data.

Process Groups

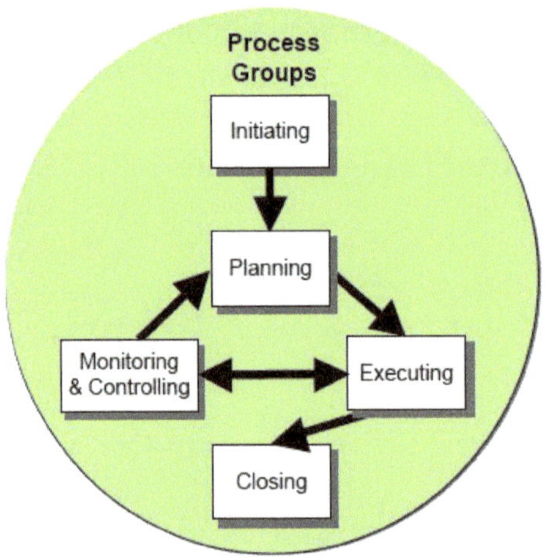

The project development stages

Traditionally (depending on what project management methodology is being used), project management includes a number of elements: four to five project management process groups, and a control system. Regardless of the methodology or terminology used, the same basic project management processes or stages of development will be used. Major process groups generally include:

- Initiation
- Planning
- Production or execution

- Monitoring and controlling
- Closing

In project environments with a significant exploratory element (e.g., research and development), these stages may be supplemented with decision points (go/no go decisions) at which the project's continuation is debated and decided. An example is the Phase–gate model.

Initiating

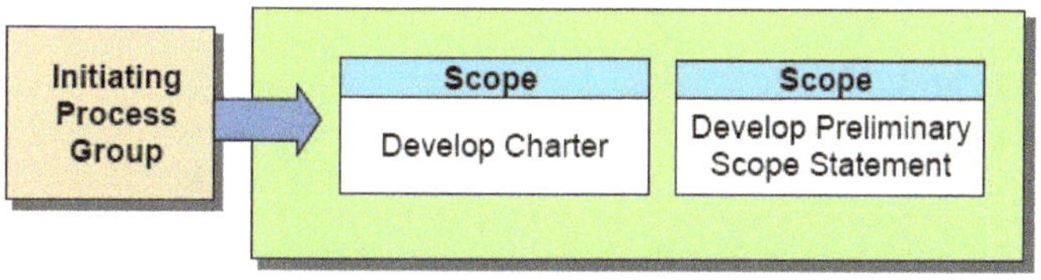

Initiating process group processes

The initiating processes determine the nature and scope of the project. If this stage is not performed well, it is unlikely that the project will be successful in meeting the business' needs. The key project controls needed here are an understanding of the business environment and making sure that all necessary controls are incorporated into the project. Any deficiencies should be reported and a recommendation should be made to fix them.

The initiating stage should include a plan that encompasses the following areas:

- analyzing the business needs/requirements in measurable goals
- reviewing of the current operations
- financial analysis of the costs and benefits including a budget
- stakeholder analysis, including users, and support personnel for the project
- project charter including costs, tasks, deliverables, and schedules
- SWOT analysis strengths, weaknesses, opportunities, and threats to the business

Planning

After the initiation stage, the project is planned to an appropriate level of detail (see example of a flow-chart). The main purpose is to plan time, cost and resources adequately to estimate the work needed and to effectively manage risk during project execution. As with the Initiation process group, a failure to adequately plan greatly reduces the project's chances of successfully accomplishing its goals.

Project planning generally consists of

- determining how to plan (e.g. by level of detail or Rolling Wave planning);
- developing the scope statement;

- selecting the planning team;
- identifying deliverables and creating the work breakdown structure;
- identifying the activities needed to complete those deliverables and networking the activities in their logical sequence;
- estimating the resource requirements for the activities;
- estimating time and cost for activities;
- developing the schedule;
- developing the budget;
- risk planning;
- developing quality assurance measures;
- gaining formal approval to begin work.

Additional processes, such as planning for communications and for scope management, identifying roles and responsibilities, determining what to purchase for the project and holding a kick-off meeting are also generally advisable.

For new product development projects, conceptual design of the operation of the final product may be performed concurrent with the project planning activities, and may help to inform the planning team when identifying deliverables and planning activities.

Executing

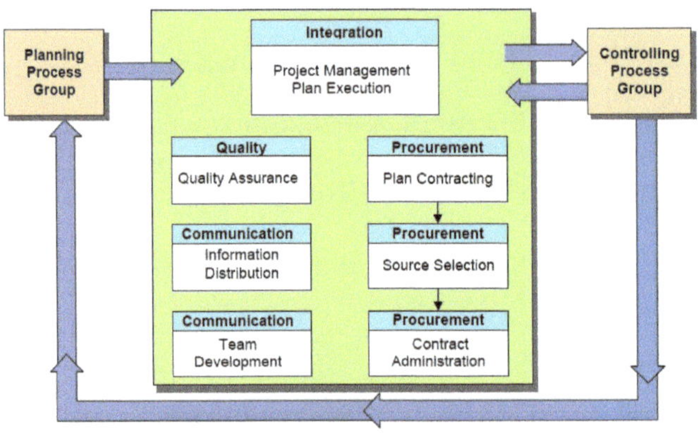

Executing process group processes

The execution/implementation phase ensures that the project management plan's deliverables are executed accordingly. This phase involves proper allocation, co-ordination and management of human resources and any other resources such as material and budgets. The output of this phase is the project deliverables.

Monitoring and Controlling

Monitoring and controlling consists of those processes performed to observe project execution so that potential problems can be identified in a timely manner and corrective action can be taken, when necessary, to control the execution of the project. The key benefit is that project performance is observed and measured regularly to identify variances from the project management plan.

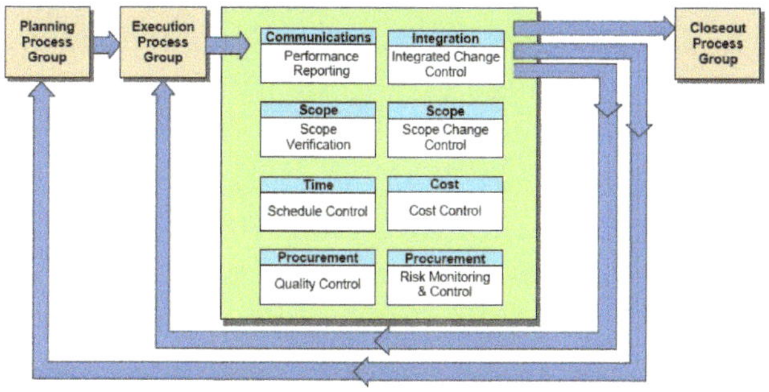

Monitoring and controlling process group processes

Monitoring and controlling includes:

- Measuring the ongoing project activities ('where we are');
- Monitoring the project variables (cost, effort, scope, etc.) against the project management plan and the project performance baseline (*where we should be*);
- Identifying corrective actions to address issues and risks properly (*How can we get on track again*);
- Influencing the factors that could circumvent integrated change control so only approved changes are implemented.

In multi-phase projects, the monitoring and control process also provides feedback between project phases, in order to implement corrective or preventive actions to bring the project into compliance with the project management plan.

Project maintenance is an ongoing process, and it includes:

- Continuing support of end-users
- Correction of errors
- Updates to the product over time

Monitoring and controlling cycle

In this stage, auditors should pay attention to how effectively and quickly user problems are resolved.

Over the course of any construction project, the work scope may change. Change is a normal and expected part of the construction process. Changes can be the result of necessary design modifications, differing site conditions, material availability, contractor-requested changes, value engineering and impacts from third parties, to name a few. Beyond executing the change in the field, the change normally needs to be documented to show what was actually constructed. This is referred to as change management. Hence, the owner usually requires a final record to show all changes or, more specifically, any change that modifies the tangible portions of the finished work. The record is made on the contract documents – usually, but not necessarily limited to, the design drawings. The end product of this effort is what the industry terms as-built drawings, or more simply, "as built." The requirement for providing them is a norm in construction contracts. Construction document management is a highly important task undertaken with the aid an online or desktop software system, or maintained through physical documentation. The increasing legality pertaining to the construction industries maintenance of correct documentation has caused the increase in the need for document management systems.

When changes are introduced to the project, the viability of the project has to be re-assessed. It is important not to lose sight of the initial goals and targets of the projects. When the changes accumulate, the forecasted result may not justify the original proposed investment in the project. Successful project management identifies these components, and tracks and monitors progress so as to stay within time and budget frames already outlined at the commencement of the project.

Closing

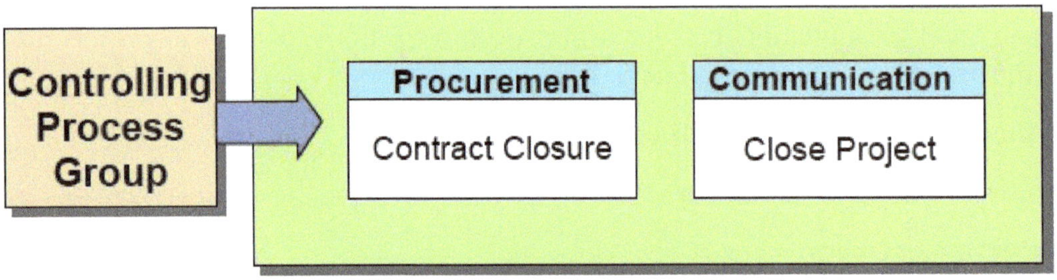

Closing process group processes

Closing includes the formal acceptance of the project and the ending thereof. Administrative activities include the archiving of the files and documenting lessons learned.

This phase consists of:

- Contract closure: Complete and settle each contract (including the resolution of any open items) and close each contract applicable to the project or project phase.
- Project close: Finalize all activities across all of the process groups to formally close the project or a project phase

Also included in this phase is the Post Implementation Review. This is a vital phase of the project for the project team to learn from experiences and apply to future projects. Normally a Post Implementation Review consists of looking at things that went well and analyzing things that went badly on the project to come up with lessons learned.

Project Controlling and Project Control Systems

Project controlling (also known as Cost Engineering) should be established as an independent function in project management. It implements verification and controlling function during the processing of a project in order to reinforce the defined performance and formal goals. The tasks of project controlling are also:

- the creation of infrastructure for the supply of the right information and its update
- the establishment of a way to communicate disparities of project parameters
- the development of project information technology based on an intranet or the determination of a project key performance indicator system (KPI)
- divergence analyses and generation of proposals for potential project regulations
- the establishment of methods to accomplish an appropriate project structure, project workflow organization, project control and governance
- creation of transparency among the project parameters

Fulfillment and implementation of these tasks can be achieved by applying specific methods and instruments of project controlling. The following methods of project controlling can be applied:

- investment analysis
- cost–benefit analysis
- value benefit analysis
- expert surveys
- simulation calculations
- risk-profile analysis
- surcharge calculations
- milestone trend analysis
- cost trend analysis
- target/actual-comparison

Project control is that element of a project that keeps it on track, on-time and within budget. Project control begins early in the project with planning and ends late in the project with post-implementation review, having a thorough involvement of each step in the process. Projects may be audited or reviewed while the project is in progress. Formal audits are generally risk or compliance-based and management will direct the objectives of the audit. An examination may include a comparison of approved project management processes with how the project is actually being managed. Each project should be assessed for the appropriate level of control needed: too much control is too time consuming, too little control is very risky. If project control is not implemented correctly, the cost to the business should be clarified in terms of errors and fixes.

Control systems are needed for cost, risk, quality, communication, time, change, procurement, and human resources. In addition, auditors should consider how important the projects are to the financial statements, how reliant the stakeholders are on controls, and how many controls exist. Auditors should review the development process and procedures for how they are implemented. The process of development and the quality of the final product may also be assessed if needed or requested. A business may want the auditing firm to be involved throughout the process to catch problems earlier on so that they can be fixed more easily. An auditor can serve as a controls consultant as part of the development team or as an independent auditor as part of an audit.

Businesses sometimes use formal systems development processes. These help assure systems are developed successfully. A formal process is more effective in creating strong controls, and auditors should review this process to confirm that it is well designed and is followed in practice. A good formal systems development plan outlines:

- A strategy to align development with the organization's broader objectives
- Standards for new systems
- Project management policies for timing and budgeting
- Procedures describing the process
- Evaluation of quality of change

Topics

Project Managers

A project manager is a professional in the field of project management. Project managers can have the responsibility of the planning, execution, controlling, and closing of any project typically relating to the construction industry, engineering, architecture, computing, and telecommunications. Many other fields of production engineering, design engineering, and heavy industrial have project managers.

A project manager is the person accountable for accomplishing the stated project objectives. Key project management responsibilities include creating clear and attainable project objectives, building the project requirements, and managing the triple constraint (now including more constraints and calling it competing constraints) for projects, which is cost, time, and scope for the first three but about three additional ones in current project management.

A project manager is often a client representative and has to determine and implement the exact needs of the client, based on knowledge of the firm they are representing. The ability to adapt to the various internal procedures of the contracting party, and to form close links with the nominated representatives, is essential in ensuring that the key issues of cost, time, quality and above all, client satisfaction, can be realized.

Project Management Types

Project management can apply to any project, but it is often tailored to accommodate the specific needs of different and highly specialized industries. For example, the construction industry, which focuses on the delivery of things like buildings, roads, and bridges, has developed its own specialized form of project management that it refers to as *Construction project management* and in which project managers can become trained and certified. The Information technology industry has also evolved to develop its own form of Project management that is referred to as *IT Project management* and which specializes in the delivery of technical assets and services that are required to pass through various lifecycle phases such as planning, design, development, testing, and deployment. *Biotechnology project management* focuses on the intricacies of biotechnology research and development.

For each type of project management, project managers develop and utilize repeatable templates that are specific to the industry they're dealing with. This allows project plans to become very thorough and highly repeatable, with the specific intent to increase quality, lower delivery costs, and lower time to deliver project results.

Risk Management

The United States Department of Defense states; "Cost, Schedule, Performance, and Risk," are the four elements through which Department of Defense acquisition professionals make trade-offs and track program status. There are also international standards. Risk management applies proactive identification of future problems and understanding of their consequences allowing predictive decisions about projects.

Work Breakdown Structure

The work breakdown structure (WBS) is a tree structure that shows a subdivision of effort required to achieve an objective—for example a program, project, and contract. The WBS may be hardware-, product-, service-, or process-oriented (see an example in a NASA reporting structure (2001)).

A WBS can be developed by starting with the end objective and successively subdividing it into manageable components in terms of size, duration, and responsibility (e.g., systems, subsystems, components, tasks, sub-tasks, and work packages), which include all steps necessary to achieve the objective.

The work breakdown structure provides a common framework for the natural development of the overall planning and control of a contract and is the basis for dividing work into definable increments from which the statement of work can be developed and technical, schedule, cost, and labor hour reporting can be established. The work breakdown structure can be displayed in two forms one in form of a table with subdivision of tasks two in form of an organisational chart.

International Standards

There have been several attempts to develop project management standards, such as:

- ISO 21500: 2012 – Guidance on project management. This is the first project management ISO.

- ISO 31000: 2009 – Risk management. Risk management is 1 of the 10 knowledge areas of either ISO 21500 or PMBoK5 concept of project management.
- ISO/IEC/IEEE 16326-2009 – Systems and Software Engineering—Life Cycle Processes—Project Management
- Capability Maturity Model from the Software Engineering Institute.
- GAPPS, Global Alliance for Project Performance Standards – an open source standard describing COMPETENCIES for project and program managers.
- A Guide to the Project Management Body of Knowledge from the Project Management Institute (PMI)
- HERMES method, Swiss general project management method, selected for use in Luxembourg and international organizations.
- The ISO standards ISO 9000, a family of standards for quality management systems, and the ISO 10006:2003, for Quality management systems and guidelines for quality management in projects.
- PRINCE2, PRojects IN Controlled Environments.
- Association for Project Management Body of Knowledge
- Team Software Process (TSP) from the Software Engineering Institute.
- Total Cost Management Framework, AACE International's Methodology for Integrated Portfolio, Program and Project Management.
- V-Model, an original systems development method.
- The Logical framework approach, which is popular in international development organizations.
- [Australian Institute of Project Management] AIPM has 4 levels of certification; CPPP, CPPM, CPPD & CPPE for Certified Practicing Project... Partner, Manager, Director and Executive.

Project Portfolio Management

An increasing number of organizations are using, what is referred to as, project portfolio management (PPM) as a means of selecting the right projects and then using project management techniques as the means for delivering the outcomes in the form of benefits to the performing private or not-for-profit organization.

Project Management Software

Project management software is software used to help plan, organize, and manage resource pools, develop resource estimates and implement plans. Depending on the sophistication of the software, functionality may include estimation and planning, scheduling, cost control and budget management, resource allocation, collaboration software, communication, decision-making, workflow, risk, quality, documentation and/or administration systems.

Virtual Project Management

Virtual program management (VPM) is management of a project done by a virtual team, though it rarely may refer to a project implementing a virtual environment It is noted that managing a virtual project is fundamentally different from managing traditional projects, combining concerns of telecommuting and global collaboration (culture, timezones, language).

Project Planning

Project planning is part of project management, which relates to the use of schedules such as Gantt charts to plan and subsequently report progress within the project environment.

Initially, the project scope is defined and the appropriate methods for completing the project are determined. Following this step, the durations for the various tasks necessary to complete the work are listed and grouped into a work breakdown structure. Project planning is often used to organize different areas of a project, including project plans, work loads and the management of teams and individuals. The logical dependencies between tasks are defined using an activity network diagram that enables identification of the critical path. Project planning is inherently uncertain as it must be done before the project is actually started. Therefore the duration of the tasks is often estimated through a weighted average of optimistic, normal, and pessimistic cases. The critical chain method adds "buffers" in the planning to anticipate potential delays in project execution. Float or slack time in the schedule can be calculated using project management software. Then the necessary resources can be estimated and costs for each activity can be allocated to each resource, giving the total project cost. At this stage, the project schedule may be optimized to achieve the appropriate balance between resource usage and project duration to comply with the project objectives. Once established and agreed, the project schedule becomes what is known as the baseline schedule. Progress will be measured against the baseline schedule throughout the life of the project. Analyzing progress compared to the baseline schedule is known as earned value management.

The inputs of the project planning phase 2 include the project charter and the concept proposal. The outputs of the project planning phase include the project requirements, the project schedule, and the project management plan.

Project planning can be done manually, but project management software is often used.

In the process of planning, alternatives are examined and the best alternative is chosen. The goal of planning is to minimize resource use (cost) while satisfactorily completing the task. Efficient use of equipments, material, labor and ensuring coordinated effort are the basic aim. The outcome of planning is predetermined course of action.Thus, the planning creates an orderly sequence of events, defines strategies to be followed in carrying forth the plan and describes ultimate disposition of the result. Putting the various activity of the project in the sequence on the time frame is the process of scheduling. Scheduling is required for continuous checking of the project (control), for resource mobilization, to minimize the cost and use of resources optimally.

Various scheduling techniques have been employed to plan the activity in sequence in project management. In construction project, bar chart and critical path method (CPM) have been widely used. During the planning process, a manager builds the facilities on paper, thus identities each of the various tasks and time. During construction, these predetermined course of action form the basis for monitoring and the checking the progress of the work. Following steps are followed during planning, scheduling and control.

a. Identifying and defining activity,
b. Defining activity interdependence,
c. Estimate time and resources for each activity,
d. Constructing the network,
e. Calculations on network for project time, earliest start and finish of activity, resource requirement, etc,
f. Project control and project review.

a. Identifying the activity: Activity is well defined task which consumes time and resources. The identification of activity depends upon the level of details and the requirement of management. Different levels of leaders requires different detail in the project planning. For example, the cost of beam can be an activity in building construction or it can be further split into different activities such as bar scheduling, making form work, concrete mixing and casting, curing the beam etc. Generally tasks to be performed in break down using 'Top-Bottom Design'. The work break down structure (WBS) forms the useful tool to identify the activity. The outcome of this phase is the complete list of activity to perform the project.

b. Defining activity interdependence: In this step, each of the activity is considered. It is argued which activity will succeed the activity under consideration. Thus, we obtain the immediate succeeding activity or immediate successor. This way of thinking is called forward planning. We can also think which activity will precede the activity under consideration and similar manner we can prepare immediate precedence list. This way of thinking is called Backward Planning Outcome of this step is list of activity together with immediate preceding or succeeding activity. This step is carried out by most knowledgeable persons of the team. Generally these steps are carried at the senior/experience persons in the department or contractor's company.

Estimation of time and resources for activity: The estimation of activity duration without bias is very important. To avoid bias in the estimate, we can follow some guidelines which are described as follows -

i. A group of experts should estimate the time and resource of the activity.
ii. The estimate of activity duration should be done activity by activity and time estimate of one activity should not affect that of the other.
iii. The estimate should be based on normal level of work for labour and equipment.
iv. In the schedule, normal days should be considered ; weekends and holidays should be specified in the project.

Methods of time estimate: Depending on the type of calculation on the schedule/network, different methods of time estimates have been in practice. The most popular methods are given below -

i. *Definite time estimation:* In this method, each activity is specifided only one time. This time is used normally to calculate project durations, finding critical path, earliest start and finish of activity, etc.

ii. *Two time estimate method:* We specify two activity duration to each of the activity. These activity durations are

   a. Normal activity duration

   b. Crash activity duration

Normal activity duration is the time of carrying out activity considering normal level of resource such as labour and equipment. Sometimes, we require to know the shortest possible activity duration, if we supply enough resources.

In this condition, cost of the activity generally goes up. The relation between cost and the duration is given as follows -

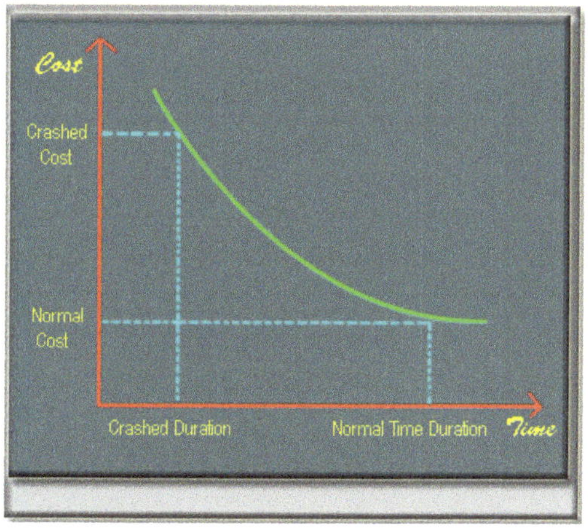

Many a time the base curve approximate as linear as shown below -

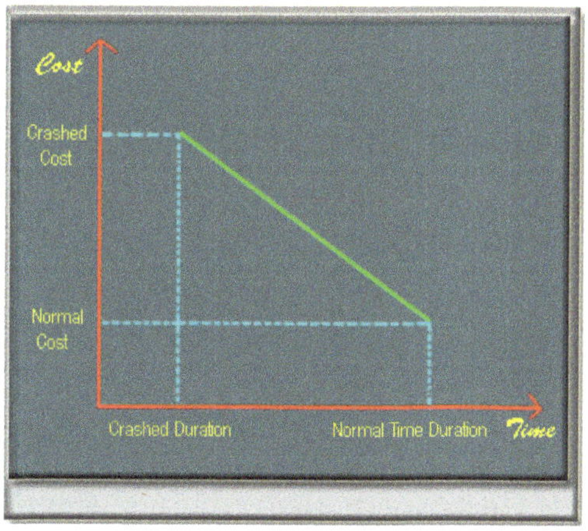

PERT time estimate: In this technique, it is assumed that the activity durations follow a stochastic duration. Generally, - distribution is assumed in the model. To specify the procedure, following three activity durations are chosen -

a. Most likely time,

b. Most optimistic time,

c. Most pessimistic time.

a. *Most likely time* (t_L): This time is the time when activity has more probability to occur than any other activity. This is the tallest pick in the probability distribution, i.e. mode.

b. *Most optimistic time* (t_o): This time has 95% fastest occurrence than the occurrence of any other activity. This is the shortest duration assumed. We mean here is that optimistic time has 19 times fastest in every 20-occurrence time.

c. *Most pessimistic time* (t_p):This time has 95% slowest occurrence than the occurrence of any other activity. This is the longest duration assumed. We mean here is that pessimistic time has 19 times slowest in every 20-occurrence time.

d. Based on the above time estimate, the average time taken for the completion of the activity or job is called expected time. Taking weighted average, we can obtain the expected time as -

$$t_E = \frac{t_o + 4t_L + t_P}{6}$$

e. This result has been derived considering $\beta-$ distribution and weighted average as one. The expected time of an activity in series can be summed to calculate the total time as -

$$\Sigma t_E = \frac{\Sigma t_o + 4\Sigma t_L + \Sigma t_P}{6}$$

Frequency distribution of activities -

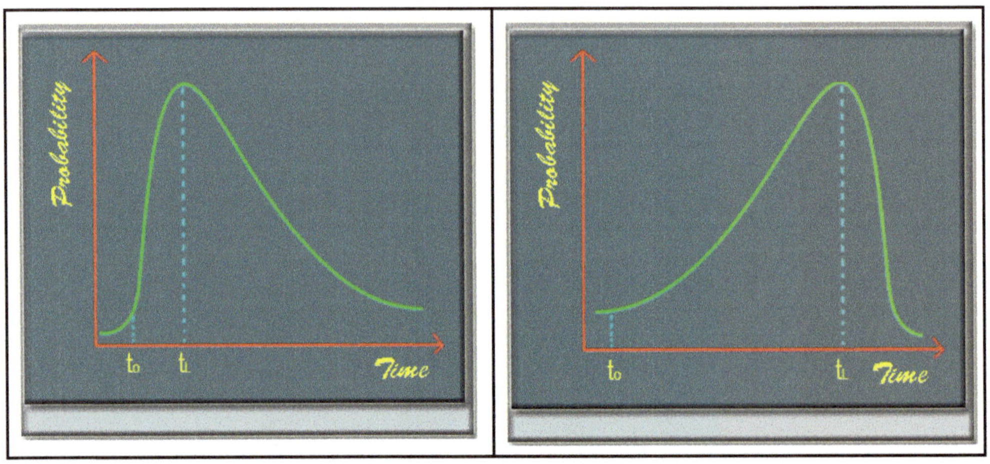

Estimation of resources: In order to estimate the resources, the project manager must consider the efficiency of the machine, productivity of labour, learning process, site conditions, impact of weather. For example, person at floor level can do more plastering than standing on the scaffolding. Similarly a person's efficiency is less at the start of the work and increases slowly and reaches the saturation level following the learning curve as shown -

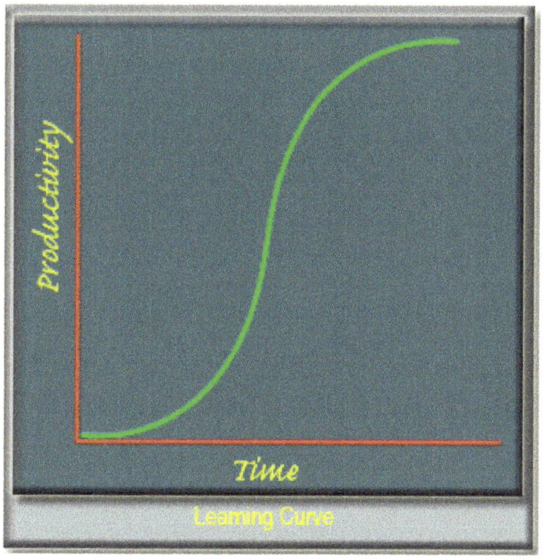

The above curve is called the learning curve. All these points should be taken into account while planning resources.

d. Construction the network: In the critical path method, network is a directed graph. A directed graph is a structure formulated by nodes, which are tied by directed arrow.

e. Let $N = \{1, 2, \ldots\ldots\ldots\ldots n\}$ be the set of nodes, and

f. $P = \{(1,2)(5,6)\ldots\ldots\ldots\ldots(I, j)\}$ be the directed arrow

g. The graph of the activity network is represented mathematically by the symbol [N,P].

h. There are two logic diagramming formats used in CPM -

 i. Activity on the arrow. (AOA)

 ii. Activity on the node. (AON)

 iii. *Activity on the arrow (AOA)*: In this format, activity is denoted by the arrow and nodes merely serve as connecting points for the activity. We often require to use activity with zero time called dummy activity to retain the logic and grammatical rules of the network. The network is constructed on the logic that an activity leaving a node cannot begin until all the activity heading into that node have been completed. This diagramming format has widely been used.

 iv. *Activity on the node (AON):* Now-a-days, CPM users employ activity on node format and many of the softwares also use this diagramming format. This format is also referred to as "Precedence Diagramming". In this format, logic in the network is represented by node and precedence arrow. The precedence arrows show the order sequence and relationship between activities. The logic behind the network diagram is that an activity cannot start until all preceding activities are complete. The logic network simply represents the relationship between the activities, provides project understanding and improves communications. The network is constructed without considering the activity duration or the resource requirement of the activity.

With the advent of computers, this stage has lost its earlier significance. Now there are many softwares available, which provide the network if succeeding activity/ preceding activity lists are fed.

e. Calculation on Network: In this stage, the earliest start, the earliest finish time, latest start time, latest finish time, critical activity, optimization for time-cost, resource planning through the network, etc, are carried out. We shall illustrate these calculations through examples.

Examples to be Given

f. Project Control and project review: The planning is pre-construction activity. The schedule of the project tells the project manager how to arrange the various activities and what resources will be required at a particular instant of time, and how much progress is to be expected. During the construction phase, the schedule serves as the yard-stick to check whether the project may be ahead of plan or behind the plan. If it is behind, then network schedule also helps for further resource mobilization.

Activity Time

In the network, it is important to know the criticality of the activity, time available to perform the activity provided date of completion of a project.

1. How early we can start the activity.
2. How much we can delay the activity such that target date of completion of project is not affected
3. What the time available to perform the activity.

To answer the above question we shall associate (define) following time with an activity

a Early start time

b) Early finish time

c) Latest start time

d) Latest finish time

Early start time: - It is the earliest time by which an activity can started.

Let t^{ij} be the activity between events i and j. Let T_E^i is the earliest expected time of the event i then Earliest start time is equal is to

$$EST = T_E^i$$

b) Earliest finish time: The earliest finish time of an activity is the time by which an activity can be finished earliest. The earliest finish time of an activity is earliest start time plus activity duration.

$$EFT = EST + \text{activity duration}$$

c) Latest finish time: The latest finish time is the time by which an activity must be finished such that completion of the project is not delayed the latest finish time of an activity between events i and j is the latest allowable event time of

$$LFT = T_L^j$$

Latest start Time: The latest start time is the time by which an activity must start such that completion of project is not delayed. The latest start time of an activity between events i and j is latest finish time minus activity duration.

$$LST = LFT - activity\ duration$$

Float:

In the last reading we have discuss that in the project with art interested in time available to execute particular task(activity) during the project period. We shall introduce the concept float which tells us how much an activity can fluctuate without affecting the completion of a project. We shall introduce following float

a) Total float.

b) Free float.

c) Independent float.

d) Interfering float.

a) Total float is the maximum time available during which an activity can fluctuate. It is maximum time available minus activity duration. The maximum time available in the difference between latest finish time minus earliest start time. Maximum time available to execute the work = LFT – EST

Hence

$total\ float = maximum\ time\ available - activity\ duration$

$= LFT - EST - t^{ij}$

$= LST - EST$

b) Free float: - Free float is the time available to fluctuate the activity without the affecting the succeeding activity. The mean of without affecting succeeding activity is that the activity must finish before the earliest occurrence of the head event.

$Time\ available = T_E^i - EST$

$Free\ float = Time\ available - activity\ duration$

$= TEi - EST - TEij$

$Free\ float = T_E^i - EFT$

c) Independent float: - It is the time available to fluctuate if the preceding activities get completed as late as possible and succeeding activities get started as early as possible.

Hence time available $= T_E^{\,i} - T_L^{\,i}$

This is the minimum time available to execute

Independent float = Minimum time available – activity duration

$$= T_E^{\,i} - T_L^{\,i} - T_E^{\,ij}$$

d) Interfering float is defined as difference between total float and free float.

Event

Event is the important point (mile stone) during execution of particular work or task. Many a time we only are concerned about commencement or completion of the task. Event is defined as start or end of an activity or activities. Thus, event is followed by more than one activity. In this case event is marked by completion of all activities or start of any one of the activity.

Example: -

1. Start of plastering is an event.
2. Completion of excavation for piping layout is an event.
3. Start of floor tiles is an event.
4. Giving power supply to fan and light.

Event is characterized by completion and start of an activity. We can defined

a) An event is said to occur if all the activity leading to it has been completed.

b) An event is said to occur if any of the activity has started.

During a project, project manager needs to know following two questions

a. How easily he can start any particular activity?
b. How late he can delay such that project does not delayed?

We shall define following two times associated with an event.

a. Earliest expected time.
b. Latest allowable occurrence time.

Earliest expect time: The earliest expected time of an event is time which tells the event can not occur earlier than this time.

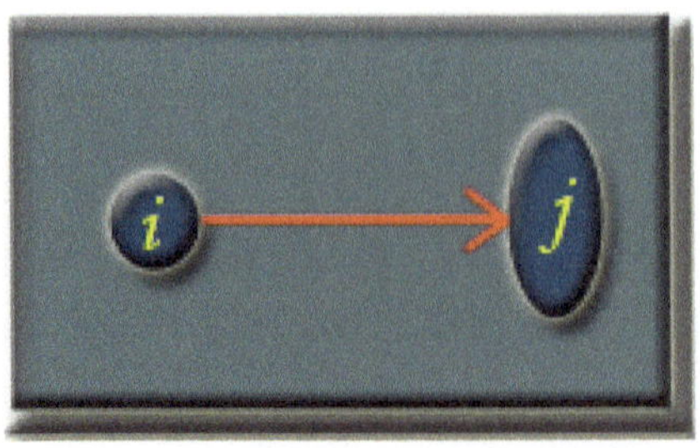

If t_{Eij} is the expected occurrences time of an activity between event i and j and T_{Ei} and T_{Ej} is the earliest occurrence time of the occurrence of this event then earliest occurrence time of occurrence of the event j is related to that of event i as

$$T_{Ej} = T_{Ei} + t_{Eij}$$

If there are many activities leading to the event j then earliest expected time of j is

$T_{Ej} = \max(T_{Ei} + t_{Eij})$ for all i

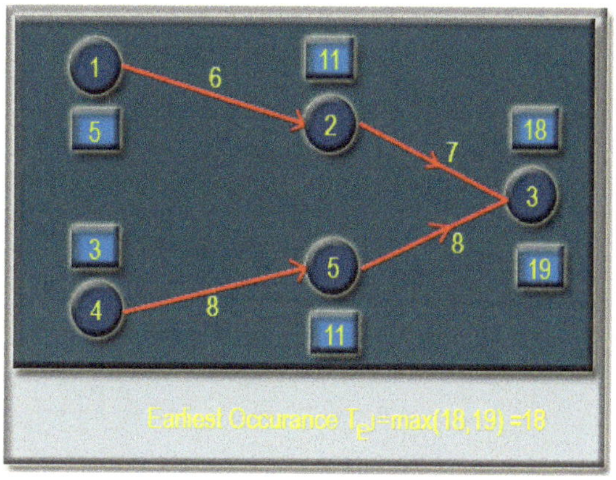

The latest allowable occurrence time is the late occurrence of the event such that schedule of the project is not disturbed. The time tells us how much the event can be delayed such that project will be completed on time.

Let us consider an event i and j connected through activity i-j as

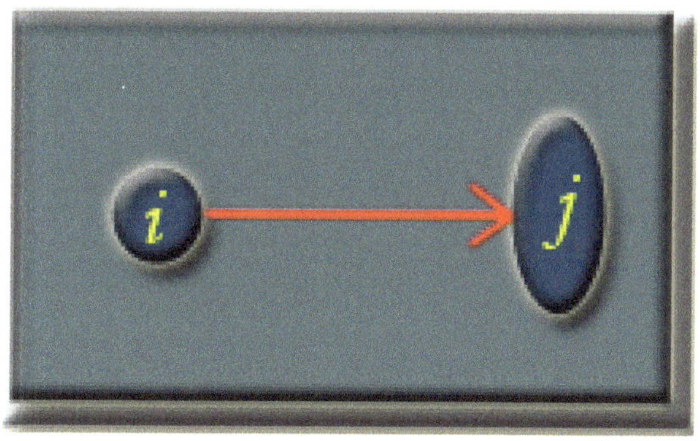

Let T_{Li} and T_{Lj} be the latest allowable occurrence time of event i and j. t_{Eij} is the expected time of activity i-j. Then the latest occurrence time of i is related to the latest occurrence time of j as

$$T_{Li} = T_{Lj} - t_{Eij}$$

Thus we can see that latest allowable occurrence time of an event has to be calculated starting from completion event of the project in backward direction. This is called Backpass.

If suppose a number of activity starts from an event then the latest allowable occurrence time is the minimum of the time calculated from all the path.

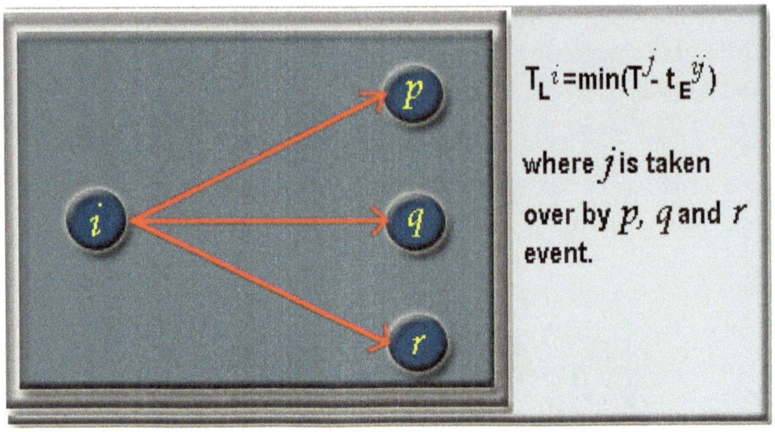

Earliest occurrence time and latest occurrence time is shown in the following figure-

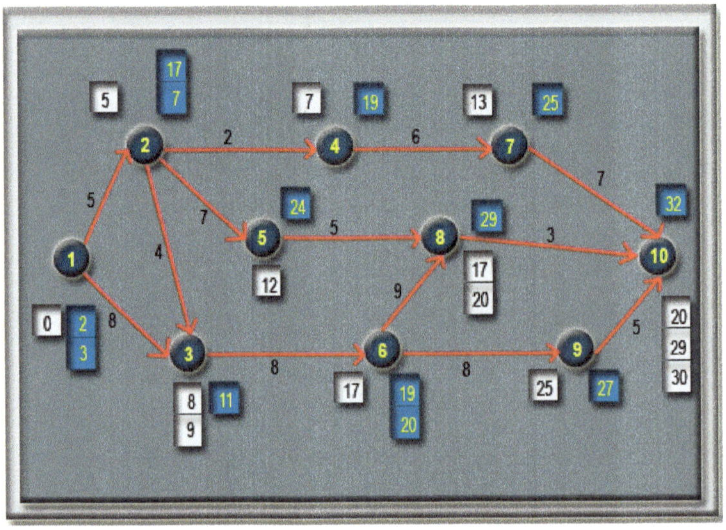

Let completion time of project is 32 days. The latest allowable occurrence time is shown in box 

The earliest expected time of an event is shown in box 

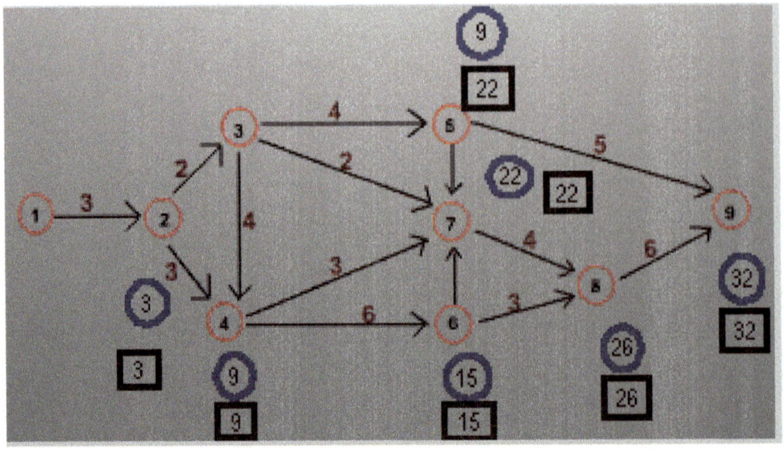

Project Management Software

Project management software has the capacity to help plan, organize, and manage resource tools and develop resource estimates. Depending on the sophistication of the software, it can manage estimation and planning, scheduling, cost control and budget management, resource allocation, collaboration software, communication, decision-making, quality management and documentation or administration systems. Today, numerous PC & browser based project management software and contract management software solutions exist, and are finding applications in almost every type of business.

History

Project Management Predecessors

The first historically relevant year for the development of Project Management Software is 1896, marked by the introduction of the Harmonogram. Polish Economist Karol Adamiecki attempted to display task development in a floating chart, and laid the foundation for project management software as it is today. 1912 was the year when the Henry Gantt replaced the Hrongram into the more advanced Gantt chart, a scheduling diagram which broke ship design tasks down for the purposes of Hoover Dam in early 1931. Today's Gantt charts are almost the same as their original counterparts, and are a critical part of all project management systems.

Emergence of the 'Project Management' Term and Modernized Techniques

The term Project Management was not used prior to 1954, when US Air Force General Bernard Adolph Schriever introduced it for military purposes. In the years to follow, project management gained relevance in the business world, a trend which had a lot to do with the formation of the American Association of Engineers AACE (1956), and Rang and DuPont's Critical Path Method calculating project duration ever since 1957.

The trend is also related to the appearance of the Program Evaluation Review Technique (PERT) in 1958. PERT went further with monitoring projects, and enabled users to monitor tasks, being at the same time able to evaluate their quality, and to estimate the time needed to accomplish each of them. The same as Gantt Charts and CPM, PERT was invented for military purposes, this time for the US Navy Polaris missile submarine program.

In 1965, there was a new improvement in project management technology. The US department of defense presented the Work Breakdown Structure (WBS) able to dissolve projects to even smaller visual units, organizing them in hierarchical tree structure. WBS was an inspiration for Winston Royce's Waterfall Method (1970) where management phases are organized in a way which doesn't allow a new task to begin before the previous ones are completed.

The First Project Management Products and Associations

In the period between 1965 and 1969 were formed two of the leading project management associations: the International Project Management Association (IPMA) in Europe, and the Project Management Institute (PMI) which trains project management professionals and issues certificates.

With business shifting towards technology-based and paperless methods, the first project management systems started to emerge. Oracle and Artemis launched their project managers in 1977, while Scitor Corporation did the same in 1979. Many improvements followed in the upcoming decades: In 1986, Carnegie Mellon University's Software Engineering Institute introduced Capability Maturity Software, a five-level project management method for rapidly maturing processes, while in 1988 users met Earned Value Management which added processes' scope and cost to the schedule. The trend continued with PRINCE2 (1996) which increased the number of processes to seven, because of which developers considered designing products for managing complex projects. In 2001, they adopted the Agile Project Management concept, and focused on adaptive planning and flexible response to changes. In 2006, users were already able to trigger Total Cost Management, a framework that helps controlling and reducing costs in project management.

SaaS and Cloud-Based Project Management Software

The SaaS (software-as-a-service) trend began in 2008, qualified by users as the most flexible type of project management software for their teams. In 2009, US News classified project management as one of the most demanded skills for obtaining a well-paid job.

From 2010 on, the most popular project management solutions were cloud-based, designed for the needs of virtual teams looking to access information from any location or device. As a result, 2012 brought the first mobile project management apps users can apply on the go.

Project Management Trends

With the advent of the Internet-of-Things, project management software was developed to incorporate testing technologies, development tools, and improved cyber security methods.

Tasks and Activities

Scheduling

One of the most common project management software tool types is scheduling tools. Scheduling tools are used to sequence project activities and assign dates and resources to them. The detail and sophistication of a schedule produced by a scheduling tool can vary considerably with the project management methodology used, the features provided and the scheduling methods supported. Scheduling tools may include support for:

- Multiple dependency relationship types between activities
- Resource assignment and leveling
- Critical path
- Activity duration estimation and probability-based simulation
- Activity cost accounting

Providing Information

Project planning software can be expected to provide information to various people or stakeholders, and can be used to measure and justify the level of effort required to complete the project(s). Typical requirements might include:

- Overview information on how long tasks will take to complete.
- Early warning of any risks to the project.
- Information on workload, for planning holidays.
- Evidence.
- Historical information on how projects have progressed, and in particular, how actual and planned performance are related.
- Optimum utilization of available resource.
- Cost maintenance.
- Collaboration with each teammates and customers.
- Instant communication to collaborators and customers.

Types

Desktop

Project management software has been implemented as a program that runs on the desktop of each user. Project management tools that are implemented as desktop software are typically single-user applications used by the project manager or another subject matter expert, such as a scheduler or risk manager..

Web-based

Project management software has been implemented as web application to be accessed using a web browser. This may also include the ability to use a smartphone or tablet to gain access to the application. Software as a Service (SaaS) is also web-based and has become a common delivery model for many business applications, including Project Management, Project Management Information System (PMIS) and Project Portfolio Management (PPM). SaaS is typically accessed by users using a thin client via a web browser.

Personal

A personal project management application is one used at home, typically to manage lifestyle or home projects. There is considerable overlap with *single user* systems, although personal project management software typically involves simpler interfaces.

Single User

A single-user system is programmed with the assumption that only one person will ever need to edit the project plan at once. This may be used in small companies, or ones where only a few people are involved in top-down project planning. Desktop applications generally fall into this category.

Collaborative

A collaborative system is designed to support multiple users modifying different sections of the plan at once; for example, updating the areas they personally are responsible for such that those estimates get integrated into the overall plan. Web-based tools, including extranets, generally fall into this category, but have the limitation that they can only be used when the user has live Internet access. To address this limitation, some software tools using client–server architecture provide a rich client that runs on users' desktop computer and replicates project and task information to other project team members through a central server when users connect periodically to the network. Some tools allow team members to check out their schedules (and others' as read only) to work on them while not on the network. When reconnecting to the database, all changes are synchronized with the other schedules.

Visual

A common problem in project management is a difficulty with both viewing and understanding large amounts of fluctuating project data. To tackle this, some project management software utilize information visualization, so that users can more easily find, analyze and make changes to their data. To avoid information overload, the visualization mantra of "overview first, zoom and filter, then details on demand" is often followed.

Construction Management

Construction project management (CM) is a professional service that uses specialized, project management techniques to oversee the planning, design, and construction of a project, from its beginning to its end. The purpose of CM is to control a project's time, cost and quality. CM is compatible with all project delivery systems, including design-bid-build, design-build, CM At-Risk and Public Private Partnerships. Every construction project features some amount of CM. However, professional construction managers, or CMs, are typically reserved for lengthy, large-scale, high budget undertakings (commercial real estate, transportation infrastructure, industrial facilities, military infrastructure, etc...), called capital projects. No matter the setting, a CM's responsibility is to the owner, and to a successful project.

The Role of a Contractor

A contractor is assigned to a construction project during the design or once the design has been completed by a licensed architect. This is done by going through a bidding process with different contractors. The contractor is selected by using one of three common selection methods: low-bid selection, best-value selection, or qualifications-based selection.

A construction manager should have the ability to handle public safety, time management, cost management, quality management, decision making, mathematics, working drawings, and human resources.

Functions

The functions of construction management typically include the following:

1. Specifying project objectives and plans including delineation of scope, budgeting, scheduling, setting performance requirements, and selecting project participants.

2. Maximizing the resource efficiency through procurement of labor, materials and equipment.

3. Implementing various operations through proper coordination and control of planning, design, estimating, contracting and construction in the entire process.

4. Developing effective communications and mechanisms for resolving conflicts.:

The Construction Management Association of America (CMAA) says the 120 most common responsibilities of a Construction Manager fall into the following 7 categories: Project Management Planning, Cost Management, Time Management, Quality Management, Contract Administration, Safety Management, and CM Professional Practice. CM professional practice includes specific activities, such as defining the responsibilities and management structure of the project management team, organizing and leading by implementing project controls, defining roles and responsibilities, developing communication protocols, and identifying elements of project design and construction likely to give rise to disputes and claims.

Seven Types of Construction

- Agricultural: Typically economical buildings, and other improvements, for agricultural purposes. Examples include barns, equipment and animal sheds, specialized fencing, storage silos and elevators, and water supply and drains such as wells, tanks, and ditches.

- Residential: Residential construction includes houses, apartments, townhouses, and other smaller, low-rise housing, small office types.

- Commercial: This refers to construction for the needs of private commerce, trade, and services. Examples include office buildings, “big box” stores, shopping centers and malls, warehouses, banks, theaters, casinos, resorts, golf courses, and larger residential structures such as high-rise hotels and condominiums.

- Institutional: This category is for the needs of government and other public organizations. Examples include schools, fire and police stations, libraries, museums, dormitories, research buildings, hospitals, transportation terminals, some military facilities, and governmental buildings.

- Industrial: Buildings and other constructed items used for storage and product production, including chemical and power plants, steel mills, oil refineries and platforms, manufacturing plants, pipelines, and seaports.

- Heavy civil: The construction of transportation infrastructure such as roads, bridges, railroads, tunnels, airports, and fortified military facilities. Dams are also included, but most other water-related infrastructure is considered environmental.

- Environmental: Environmental construction was part of heavy civil, but is now separate, dealing with projects that improve the environment. Some examples are water and wastewater treatment plants, sanitary and storm sewers, solid waste management, and air pollution control.

Jobs

- Architect
- Architect-Engineer
- Civil engineer
- Construction manager
- Design engineer
- Design manager
- Estimator
- Field engineer
- Planning engineer
- Project Architect
- Project co-ordinator
- Project document controller
- Project engineer
- Project manager
- Quantity surveyor
- Scheduler/scheduling engineer

Obtaining the Project

Bids

A bid is given to the owner by construction managers that are willing to complete their construction project. A bid tells the owner how much money they should expect to pay the construction management company in order for them to complete the project.

- Open bid: An open bid is used for public projects. Any and all contractors are allowed to submit their bid due to public advertising.
- Closed bid: A closed bid is used for private projects. A selection of contractors are sent an invitation for bid so only they can submit a bid for the specified project.

Selection Methods

- Low-bid selection: This selection focuses on the price of a project. Multiple construction management companies submit a bid to the owner that is the lowest amount they are willing to do the job for. Then the owner usually chooses the company with the lowest bid to complete the job for them.

- Best-value selection: This selection is focuses on both the price and qualifications of the contractors submitting bids. This means that the owner chooses the contractor with the best price and the best qualifications. The owner decides by using a request for proposal (RFP), which provides the owner with the contractor's exact form of scheduling and budgeting that the contractor expects to use for the project.
- Qualifications-based selection: This selection is used when the owner decides to choose the contractor only on the basis of their qualifications. The owner then uses a request for qualifications (RFQ), which provides the owner with the contractor's experience, management plans, project organization, and budget and schedule performance. The owner may also ask for safety records and individual credentials of their members. This method is most often used when the contractor is hired early during the design process so that the contractor can provide input and cost estimates as the design develops.

Payment Contracts

- Lump sum: This is the most common type of contract. The construction manager and the owner agree on the overall cost of the construction project and the owner is responsible for paying that amount whether the construction project exceeds or falls below the agreed price of payment.
- Cost plus fee: This contract provides payment for the contractor including the total cost of the project as well as a fixed fee or percentage of the total cost. This contract is beneficial to the contractor since any additional costs will be paid for, even though they were unexpected for the owner.
- Guaranteed maximum price: This contract is the same as the cost-plus-fee contract although there is a set price that the overall cost and fee do not go above.
- Unitprice: This contract is used when the cost cannot be determined ahead of time. The owner provides materials with a specific unit price to limit spending.

Project Stages

Design

The design stage involves four steps: programming and feasibility, schematic design, design development, and contract documents. It is the responsibility of the design team to ensure that the design meets all building codes and regulations. It is during the design stage that the bidding process takes place.

- Conceptual/Programming and feasibility: The needs, goals, and objectives must be determined for the building. Decisions must be made on the building size, number of rooms, how the space will be used, and who will be using the space. This must all be considered to begin the actual designing of the building. This phase is normally a written list of each room or space, the critical information about those spaces, and the approximate square footage of each area.
- Schematic design: Schematic designs are sketches used to identify spaces, shapes, and pat-

terns. Materials, sizes, colors, and textures must be considered in the sketches. This phase usually involves developing the floor plan, elevations, a site plan, and possibly a few details.

- Design development (DD): This step requires research and investigation into what materials and equipment will be used as well as their cost. During this phase, the drawings are refined with information from structural, plumbing, mechanical, and electrical engineers. It also involves a more rigorous evaluation how the applicable building codes will impact the project.
- Contract documents (CDs): Contract documents are the final drawings and specifications of the construction project. They are used by contractors to determine their bid while builders use them for the construction process. Contract documents can also be called working drawings.

Pre-construction

The pre-construction stage begins when the owner gives a notice to proceed to the contractor that they have chosen through the bidding process. A notice to proceed is when the owner gives permission to the contractor to begin their work on the project. The first step is to assign the project team which includes the project manager (PM), contract administrator, superintendent, and field engineer.

- Project manager: The project manager is in charge of the project team.
- Contract administrator: The contract administrator assists the project manager as well as the superintendent with the details of the construction contract.
- Superintendent: It is the superintendent's job to make sure everything is on schedule including flow of materials, deliveries, and equipment. They are also in charge of coordinating on-site construction activities.
- Field engineer: A field engineer is considered an entry-level position and is responsible for paperwork.

During the pre-construction stage, a site investigation must take place. A site investigation takes place to discover if any steps need to be implemented on the job site. This is in order to get the site ready before the actual construction begins. This also includes any unforeseen conditions such as historical artifacts or environment problems. A soil test must be done to determine if the soil is in good condition to be built upon.

Procurement

The procurement stage is when labor, materials and equipment needed to complete the project are purchased. This can be done by the general contractor if the company does all their own construction work. If the contractor does not do their own work, they obtain it through subcontractors. Subcontractors are contractors who specialize in one particular aspect of the construction work such as concrete, welding, glass, or carpentry. Subcontractors are hired the same way a general contractor would be, which is through the bidding process. Purchase orders are also part of the procurement stage.

- Purchase orders: A purchase order is used in various types of businesses. In this case, a purchase order is an agreement between a buyer and seller that the products purchased meet the required specifications for the agreed price.

Construction

The construction stage begins with a pre-construction meeting brought together by the superintendent. The pre-construction meeting is meant to make decisions dealing with work hours, material storage, quality control, and site access. The next step is to move everything onto the construction site and set it all up.

A Contractor progress payment schedule is a schedule of when (according to project milestones or specified dates) contractors and suppliers will be paid for the current progress of installed work.

Progress payments are partial payments for work completed during a portion, usually a month, during a construction period. Progress payments are made to general contractors, subcontractors, and suppliers as construction projects progress. Payments are typically made on a monthly basis but could be modified to meet certain milestones. Progress payments are an important part of contract administration for the contractor. Proper preparation of the information necessary for payment processing can help the contractor financially complete the project.

Owner Occupancy

Once the owner moves into the building, a warranty period begins. This is to ensure that all materials, equipment, and quality meet the expectations of the owner that are included within the contract.

Issues Resulting from Construction

Dust and Mud

When construction vehicles are driving around a site or moving earth, a lot of dust is created, especially during the dryer months. This may cause disruption for surrounding businesses or homes. A popular method of dust control is to have a water truck driving through the site spraying water on the dry dirt to minimize the movement of dust within and out of the construction site. When water is introduced mud is created. This mud sticks to the tires of the construction vehicles and is often lead out to the surrounding roads. A good practice is to have a street sweeper clean the roads at least once a day to minimize dirty road conditions. April 2015

Environmental Protections

- Storm water pollution: As a result of construction, the soil is displaced from its original location which can possibly cause environmental problems in the future. Runoff can occur during storms which can possibly transfer harmful pollutants through the soil to rivers, lakes, wetlands, and coastal waters.
- Endangered species: If endangered species have been found on the construction site, the site must be shut down for some time. The construction site must be shut down for as long

as it takes for authorities to make a decision on the situation. Once the situation has been assessed, the contractor makes the appropriate accommodations to not disturb the species.

- Vegetation: There may often be particular trees or other vegetation that must be protected on the job site. This may require fences or security tape to warn builders that they must not be harmed.
- Wetlands: The contractor must make accommodations so that erosion and water flow are not affected by construction. Any liquid spills must be maintained due to contaminants that may enter the wetland.
- Historical or cultural artifacts: Artifacts may include arrowheads, pottery shards, and bones. All work comes to a halt if any artifacts are found and will not resume until they can be properly examined and removed from the area.

Construction Activity Documentation

Project meetings take place at scheduled intervals to discuss the progress on the construction site and any concerns or issues. The discussion and any decisions made at the meeting must be documented.

Diaries, logs, and daily field reports keep track of the daily activities on a job site each day.

- Diaries: Each member of the project team is expected to keep a project diary. The diary contains summaries of the day's events in the member's own words. They are used to keep track of any daily work activity, conversations, observations, or any other relevant information regarding the construction activities. Diaries can be referred to when disputes arise and a diary happens to contain information connected with the disagreement. Diaries that are handwritten can be used as evidence in court.
- Logs: Logs keep track of the regular activities on the job site such as phone logs, transmittal logs, delivery logs, and RFI (Request for Information) logs.
- Daily field reports: Daily field reports are a more formal way of recording information on the job site. They contain information that includes the day's activities, temperature and weather conditions, delivered equipment or materials, visitors on the site, and equipment used that day.

Labor statements are required on a daily basis. Also list of Labor, PERT CPM are needed for labor planning to complete a project in time.

Resolving Disputes

- Mediation: Mediation uses a third party mediator to resolve any disputes. The mediator helps both disputing parties to come to a mutual agreement. This process ensures that no attorneys become involved in the dispute and is less time-consuming.
- Minitrial: A minitrial takes more time and money than a mediation. The minitrial takes place in an informal setting and involves some type of advisor or attorney that must be

paid. The disputing parties may come to an agreement or the third party advisor may offer their advice. The agreement is nonbinding and can be broken.

- Arbitration: Arbitration is the most costly and time-consuming way to resolve a dispute. Each party is represented by an attorney while witnesses and evidence are presented. Once all information is provided on the issue, the arbitrator makes a ruling which provides the final decision. The arbitrator provides the final decision on what must be done and it is a binding agreement between each of the disputing parties.

Terminology

Berlin Brandenburg Airport, an example of poor construction management

The following terms are commonly used in the industry:

- Construction management (CM) refers to form of delivery and (in the UK) management of the site.
- Real estate management (REM) is professional property advice (a continuous process, as opposed to a process).
    - Corporate real estate management (CREM) is a company-focused variant.
- Management contracting (MC):
    - UK: form of delivery
    - US: CM at risk
- Programme management (ProgM) is concerned with management of a client's portfolio (the programme, in this sense, is equivalent to a client's brief) or (in the UK) managing time in a project.
- Project control (PC) is the tracking and reporting of the progress, time, cost and quality of a project. This function can be characterized as passive, whereas construction project management (CPM) is active.
- Project leader (PL): The person responsible for achieving the project's objectives; acts as a manager "in-line".

- Project director (PD): The leader of a large project that can be broken down into sub-projects (e.g. the Channel tunnel) or the head of a PM organization
- Owner representative (OR): The representative of the owner; may be internal or external to the company
- Document Control (DC): Key function of a project manager
- Build operate transfer (BOT)
    - Finance build operate transfer (FBOT)
    - Design build operate transfer (DBOT)
- Build own operate (BOO)
- Engineering, procurement, and construction (EPC)
- Private finance initiative (PFI)
- General contract (GC)
- Joint Venture (JV): A collaboration between two or more companies from the same or different backgrounds and/or fields to complete a common project. Joint Ventures typically have a lead contractor that deals with most of the business aspects. The lead will usually have a bigger stake in the partnership (i.e. the lead will have a 65% stake while the second contractor might have 30% and a third might have 5%).
- Guaranteed maximum price (GMP): A contract-specified upper limit to the project's final price
- Multiple prime contracts (MPC)
    - UK: One contractor takes responsibility for the development (package deal).
    - US: A client may have five or six prime contractors.

Study and Practice

Construction Management education comes in a variety of formats: formal degree programs (Two-year associate degree; four-year baccalaureate degree, masters degree, project management, operations management engineer degree, doctor of philosophy degree, postdoctoral researcher); on-the-job-training; and continuing education and professional development. Information on degree programs is available from ABET, the American Council for Construction Education (ACCE), the Construction Management Association of America (CMAA) or the Associated Schools of Construction (ASC).

According to the American Council for Construction Education (one of the academic accreditation agencies responsible for accrediting construction management programs in the U.S.), the academic field of construction management encompasses a wide range of topics. These range from general management skills, through management skills specifically related to construction, to technical knowledge of construction methods and practices. There are many schools offering Construction Management programs, including some offering a master's degree. Many ACCE Accredited Programs offer the American Institute of Constructors, Associate Constructor certification examination as a means to assess their students knowledge and skills.

Software

Capital project management software (CPMS) refers to the systems that are currently available that help capital project owner/operators, program managers, and construction managers, control and manage the vast amount of information that capital construction projects create. A collection, or portfolio of projects only makes this a bigger challenge. These systems go by different names: capital project management software, computer construction software, construction management software, project management information systems. Usually Construction Management can be referred as subset of CPMS where the scope of CPMS is not limited to construction phases of project. Among main construction management software can be mentioned Procore, PlanGrid etc...

Required Knowledge

- Construction and building
- Technology
- Public safety
- Customer service
- Human resources
- Mathematics

Skills and Abilities

- Time management
- Negotiation
- Decision making
- Problem solving

Business Model

The construction industry typically includes three parties: an owner, a licensed designer (architect or engineer) and a builder (usually known as a general contractor). There are traditionally two contracts between these parties as they work together to plan, design and construct the project. The first contract is the owner-designer contract, which involves planning, design, and construction contract administration. The second contract is the owner-contractor contract, which involves construction. An indirect third-party relationship exists between the designer and the contractor, due to these two contracts.

Skyscrapers under construction in Panama City, Panama

An owner may also contract with a construction project management company as an adviser, creating a third contract relationship in the project. The construction manager's role is to provide construction advice to the designer, design advice to the constructor on the owner's behalf and other advice as necessary.

Design, Bid, Build Contracts

The phrase "design, bid, build" describes the prevailing model of construction management, in which the general contractor is engaged through a tender process after designs have been completed by the architect or engineer.

Design-build Contracts

Many owners – particularly government agencies – let out contracts known as *design-build contracts*. In this type of contract, the construction team (known as the design-builder) is responsible for taking the owner's concept and completing a detailed design before (following the owner's approval of the design) proceeding with construction. Virtual design and construction technology may be used by contractors to maintain a tight construction time.

There are three main advantages to a design-build contract. First, the construction team is motivated to work with the architect to develop a practical design. The team can find creative ways to reduce construction costs without reducing the function of the final product. The second major advantage involves the schedule. Many projects are commissioned within a tight time frame. Under a traditional contract, construction cannot begin until after the design is finished and the project has been awarded to a bidder. In a design-build contract the contractor is established at the outset, and construction activities can proceed concurrently with the design. The third major advantage is that the design-build contractor has an incentive to keep the combined design and construction costs within the owner's budget. If speed is important, design and construction contracts can be awarded separately; bidding takes place on preliminary plans in a not-to-exceed contract instead of a single, firm design-build contract.

The major problem with design-build contracts is an inherent conflict of interest. In a standard contract the architect works for the owner and is directly responsible to the owner. In design-build the architect works for the design-builder, not the owner, therefor the design-builder may make design and construction decisions that benefit the design-builder, but that do not benefit the owner. During construction, the architect normally acts as the owner's representative. This includes reviewing the builder's work and ensuring that the products and methods meet specifications and codes. The architect's role is compromised when the architect works for the design-builder and not for the owner directly. Thus, the owner may get a building that is over-designed to increase profits for the design-builder, or a building built with lesser-quality products to maximize profits.

Planning and Scheduling

Project-management methodology is as follows:

- Work breakdown structure
- Project network of activities
  - Critical path method (CPM)
  - Resource management
  - Resource leveling
  - Risk assessment

Architecture–engineer

- Work inspection
- Change orders
- Review payments
- Materials and samples
- Shop drawings
- Three-dimensional image

Agency CM

Construction cost management is a fee-based service in which the construction manager (CM) is responsible exclusively to the owner, acting in the owner's interests at every stage of the project. The construction manager offers impartial advice on matters such as:

- Optimum use of available funds
- Control of the scope of the work
- Project scheduling

- Optimum use of design and construction firms' skills and talents
- Avoidance of delays, changes and disputes
- Enhancing project design and construction quality
- Optimum flexibility in contracting and procurement
- Cash-flow management

Comprehensive management of every stage of the project, beginning with the original concept and project definition, yields the greatest benefit to owners. As time progresses beyond the pre-design phase, the CM's ability to effect cost savings diminishes. The agency CM can represent the owner by helping select the design and construction teams and managing the design (preventing scope creep), helping the owner stay within a predetermined budget with value engineering, cost-benefit analysis and best-value comparisons. The software-application field of construction collaboration technology has been developed to apply information technology to construction management.

CM at-risk

CM at-risk is a delivery method which entails a commitment by the construction manager to deliver the project within a Guaranteed Maximum Price (GMP). The construction manager acts as a consultant to the owner in the development and design phases (preconstruction services), and as a general contractor during construction. When a construction manager is bound to a GMP, the fundamental character of the relationship is changed. In addition to acting in the owner's interest, the construction manager must control construction costs to stay within the GMP.

CM at-risk is a global term referring to the business relationship of a construction contractor, owner and architect (or designer). Typically, a CM at-risk arrangement eliminates a "low-bid" construction project. A GMP agreement is a typical part of the CM-and-owner agreement (comparable to a "low-bid" contract), but with adjustments in responsibility for the CM. The advantage of a CM at-risk arrangement is budget management. Before a project's design is completed (six to eighteen months of coordination between designer and owner), the CM is involved with estimating the cost of constructing a project based on the goals of the designer and owner (design concept) and the project's scope. In balancing the costs, schedule, quality and scope of the project, the design may be modified instead of redesigned; if the owner decides to expand the project, adjustments can be made before pricing. To manage the budget before design is complete and construction crews mobilized, the CM conducts site management and purchases major items to efficiently manage time and cost.

Advantages

- CM is working "at risk", therefore have incentive to act in the owner's interest, as well as to efficiently manage construction costs, considering they would be liable for any amount in excess of the GMP
- Ability to handle changes in design or scope

Drawbacks

- If a cost overrun occurred, it could cost the CM a great deal of money

- The CM is allowed some mistake-related contingency, therefore there is a possibility that they will compensate by reducing the scope of the work to fit the GMP
- Since the GMP is settled before design begins, it is difficult for owners to know whether they received the best possible bid

Bottom Line

An At-Risk delivery method is best for large project—both complete construction and renovation—that are not easy to define, have a possibility of changing in scope, or have strict schedule deadlines. Additionally, it is an efficient method in projects containing technical complexity, multi-trade coordination, or multiple phases.

Accelerated Construction Techniques

Starting with its Accelerated Bridge Program in the late 2000s, the Massachusetts Department of Transportation began employing accelerated construction techniques, in which it signs contracts with incentives for early completion and penalties for late completion, and uses intense construction during longer periods of complete closure to shorten the overall project duration and reduce cost.

References

- Edward R. Marsh (1975). "The Harmonogram of Karol Adamiecki". In: The Academy of Management Journal. Vol. 18, No. 2 (Jun., 1975), p. 358. (online)
- *The Definitive Guide to Project Management. Nokes, Sebastian. 2nd Ed.n. London (Financial Times / Prentice Hall): 2007. ISBN 978-0-273-71097-4
- "Standards and criteria for accreditation of post-secondary construction education degree programs" (PDF). American Council for Construction Management. Retrieved 2006-05-29
- Serra, C. E. M.; Kunc, M. (2014). "Benefits Realisation Management and its influence on project success and on the execution of business strategies". International Journal of Project Management. 33 (1): 53–66. doi:10.1016/j.ijproman.2014.03.011
- Paul C. Dinsmore et al (2005) The right projects done right! John Wiley and Sons, 2005. ISBN 0-7879-7113-8. p.35 and further
- Stagner, Steve. "Design-Build and Alternative Project Delivery in Texas" (PDF). Texas Council of Engineering Companies. Retrieved 18 October 2011
- Antunes, Ricardo; Gonzalez, Vicente (2015-03-03). "A Production Model for Construction: A Theoretical Framework". Buildings. 5 (1): 209–228. doi:10.3390/buildings5010209
- Cattani, G., Ferriani, S., Frederiksen, L. and Florian, T. (2011) Project-Based Organizing and Strategic Management, Advances in Strategic Management, Vol 28, Emerald, ISBN 1780521936

Permissions

All chapters in this book are published with permission under the Creative Commons Attribution Share Alike License or equivalent. Every chapter published in this book has been scrutinized by our experts. Their significance has been extensively debated. The topics covered herein carry significant information for a comprehensive understanding. They may even be implemented as practical applications or may be referred to as a beginning point for further studies.

We would like to thank the editorial team for lending their expertise to make the book truly unique. They have played a crucial role in the development of this book. Without their invaluable contributions this book wouldn't have been possible. They have made vital efforts to compile up to date information on the varied aspects of this subject to make this book a valuable addition to the collection of many professionals and students.

This book was conceptualized with the vision of imparting up-to-date and integrated information in this field. To ensure the same, a matchless editorial board was set up. Every individual on the board went through rigorous rounds of assessment to prove their worth. After which they invested a large part of their time researching and compiling the most relevant data for our readers.

The editorial board has been involved in producing this book since its inception. They have spent rigorous hours researching and exploring the diverse topics which have resulted in the successful publishing of this book. They have passed on their knowledge of decades through this book. To expedite this challenging task, the publisher supported the team at every step. A small team of assistant editors was also appointed to further simplify the editing procedure and attain best results for the readers.

Apart from the editorial board, the designing team has also invested a significant amount of their time in understanding the subject and creating the most relevant covers. They scrutinized every image to scout for the most suitable representation of the subject and create an appropriate cover for the book.

The publishing team has been an ardent support to the editorial, designing and production team. Their endless efforts to recruit the best for this project, has resulted in the accomplishment of this book. They are a veteran in the field of academics and their pool of knowledge is as vast as their experience in printing. Their expertise and guidance has proved useful at every step. Their uncompromising quality standards have made this book an exceptional effort. Their encouragement from time to time has been an inspiration for everyone.

The publisher and the editorial board hope that this book will prove to be a valuable piece of knowledge for students, practitioners and scholars across the globe.

Index